AF576060

BIM-Based Life Cycle Sustainability Assessment for Buildings

BIM-Based Life Cycle Sustainability Assessment for Buildings

Editor

Antonio Garcia-Martinez

MDPI • Basel • Beijing • Wuhan • Barcelona • Belgrade • Manchester • Tokyo • Cluj • Tianjin

Editor
Antonio Garcia-Martinez
Universidad de Sevilla
Spain

Editorial Office
MDPI
St. Alban-Anlage 66
4052 Basel, Switzerland

This is a reprint of articles from the Special Issue published online in the open access journal *Sustainability* (ISSN 2071-1050) (available at: https://www.mdpi.com/journal/sustainability/special_issues/buildings_assessment).

For citation purposes, cite each article independently as indicated on the article page online and as indicated below:

LastName, A.A.; LastName, B.B.; LastName, C.C. Article Title. *Journal Name* **Year**, *Volume Number*, Page Range.

ISBN 978-3-0365-5519-5 (Hbk)
ISBN 978-3-0365-5520-1 (PDF)

Contents

 *sustainability*

Editorial

BIM-Based Life Cycle Sustainability Assessment for Buildings

Antonio Garcia-Martinez

Research Group TEP 130, Architecture, Heritage and Sustainability: Acoustics, Lighting, Optics and Energy Department of Architectural Construction, Research, University Institute of Architecture and Construction Sciences IUACC, Universidad de Sevilla, 41012 Seville, Spain; agarcia6@us.es

Citation: Garcia-Martinez, A. BIM-Based Life Cycle Sustainability Assessment for Buildings. *Sustainability* **2022**, *14*, 11902. https://doi.org/10.3390/su141911902

Received: 7 September 2022
Accepted: 13 September 2022
Published: 21 September 2022

The construction of buildings has a high level of environmental impacts. Life Cycle Analysis (LCA) has been configured as an effective tool to anticipate, evaluate, and optimize these impacts. The conventional application of this methodology in the field of building construction involves the consumption of a large amount of time and resources. The recent development and progress in the integration of digital tools such as Building Information Modeling (BIM) in the LCA methodology is generating important advances in the process of optimizing environmental impacts in the buildings sector. This Special Issue, "BIM-Based Life Cycle Sustainability Assessment for Buildings", gathers some of the advances that are currently taking place in the integration of Building Information Modeling platforms in the process of minimizing the impacts that buildings cause throughout their entire life cycle.

This Special Issue covers three important areas of study: (1) workflows in LCA calculation procedures based on BIM platforms; (2) the automation of Building Assessment Analysis processes via the integration of LCA and BIM; and (3) the implementation of BIM platforms for the life cycle management of buildings.

Regarding workflow issues in calculating LCA from BIM, two papers (contributions 4 and 5) deal with the procedures used when linking BIM platforms with LCA. The paper by Regitze Kjær Zimmermann, Simone Bruhn and Harpa Birgisdóttir (contribution 4) investigates the needs and practices of integration between BIM and LCA in the building sector. This paper analyzes the BIM–LCA workflows of eight companies that have integrated LCA into BIM, identifies the data used for the BIM–LCA integration, and compiles the main challenges facing this integration. Sungwoo Lee, Sungho Tae, Hyungjae Jang, Chang U. Chae and Youngjin Bok (contribution 5) propose a method of the practical integration of Life Cycle Inventory calculation from the elaboration of BIM libraries and templates.

Advances in the automation of Building Sustainability Analysis (BSA) processes from the integration of LCA and BIM are covered by two investigations (contributions 2 and 4), which propose different procedures to generate building evaluations from two different geographical perspectives: South and Central Europe. Jan Růžička, Jakub Veselka, Zdeněk Rudovský, Stanislav Vitásek and Petr Hájek (contribution 2) describe a BIM–BSA–LCA data workflow for automatic assessment based on the experience gained on a case study of a residential building. The building quality was tested using SBToolCZ, the Czech national assessment method. José Pedro Carvalho, Ismael Alecrim, Luís Bragança and Ricardo Mateus (contribution 4) address the relationship between BIM, BSA and LCA by performing an LCA for a Portuguese case study. A set of sustainability criteria from SBTool were assessed simultaneously during the process.

Concerning the implementation of BIM platforms for the management of the life cycle of buildings, four papers (contributions 1, 3, 7 and 8) cover various phases of the building life cycle from different perspectives. Manuel Castellano-Román, Antonio Garcia-Martinez and María Luisa Pérez López (contribution 1) analyze and evaluate the maintenance and management workflow of social housing. To do so, they take the case study of AVRA, one of the public companies that manages more than 70,000 homes, proposing a BIM-based life cycle management workflow. Mochamad Agung Wibowo, Naniek Utami Handayani and Anita Mustikasari (contribution 3) propose a reverse logistics model for the construction industry, incorporating the dimensions, elements and indicators needed for the evaluation of the reverse logistics configuration. Nawal Abdunasseer Hmidah, Nuzul Azam Haron, Aidi Hizami Alias, Teik Hua Law, Abubaker Basheer Abdalwhab Altohami and Raja Ahmad Azmeer Raja Ahmad Effendi (contribution 7) review targets of the BIM interface, the BIM multi-model approach and the role of employing algorithms in BIM optimization to introduce the need for automation in the BIM technique. Abubaker Basheer Abdalwhab Altohami, Nuzul Azam Haron, Aidi Hizami Ales@Alias and Teik Hua Law (contribution 8) provide a comprehensive review that explores and identifies common emerging areas of application and common design patterns of traditional BIM–IoT integration, followed by devising better methodologies to integrate IoT into BIM.

To summarize, various areas are covered in this Special Issue. We hope that the contributions gathered in this Special Issue can offer solutions and inspire new research in the field of integrating Life Cycle Analysis methods and Building Information Modeling platforms.

List of Contributions

1. Castellano-Román, M.; Garcia-Martinez, A.; Pérez López, M.L. Social Housing Life Cycle Management: Workflow for the Enhancement of Digital Management Based on Building Information Modelling (BIM). *Sustainability* **2022**, *14*, 7488.
2. Růžička, J.; Veselka, J.; Rudovský, Z.; Vitásek, S.; Hájek, P. BIM and Automation in Complex Building Assessment. *Sustainability* **2022**, *14*, 2237.
3. Wibowo, M.A.; Handayani, N.U.; Mustikasari, A.; Wardani, S.A.; Tjahjono, B. Re-verse Logistics Performance Indicators for the Construction Sector: A Building Project Case. *Sustainability* **2022**, *14*, 963.
4. Carvalho, J.P.; Alecrim, I.; Bragança, L.; Mateus, R. Integrating BIM-Based LCA and Building Sustainability Assessment. *Sustainability* **2020**, *12*, 7468.
5. Lee, S.; Tae, S.; Jang, H.; Chae, C.U.; Bok, Y. Development of Building Information Modeling Template for Environmental Impact Assessment. *Sustainability* **2021**, *13*, 3092.
6. Zimmermann, R.K.; Bruhn, S.; Birgisdóttir, H. BIM-Based Life Cycle Assessment of Buildings—An Investigation of Industry Practice and Needs. *Sustainability* **2021**, *13*, 5455.
7. Hmidah, N.A.; Haron, N.A.; Alias, A.H.; Law, T.H.; Altohami, A.B.A.; Effendi, R.A.A.R.A. The Role of the Interface and Interface Management in the Optimization of BIM Multi-Model Applications: A Review. *Sustainability* **2022**, *14*, 1869.
8. Altohami, A.B.A.; Harun, N.A.; Ales Alias, A.H.; Law, T.H. Investigating Ap-proaches of Integrating BIM, IoT, and Facility Management for Renovating Existing Buildings: A Review. *Sustainability* **2021**, *13*, 3930.

Conflicts of Interest: The author declares no conflict of interest.

Article

Social Housing Life Cycle Management: Workflow for the Enhancement of Digital Management Based on Building Information Modelling (BIM)

Manuel Castellano-Román [1], Antonio Garcia-Martinez [2,*] and María Luisa Pérez López [3]

[1] Research Group HUM 799, Heritage Knowledge Strategies, Department of Architectural Graphic Expression, University Institute of Architecture and Construction Sciences IUACC, Universidad de Sevilla, 41012 Seville, Spain; manuelcr@us.es

[2] Research Group TEP 130, Architecture, Heritage and Sustainability: Acoustics, Lighting, Optics and Energy Department of Architectural Construction, Research, University Institute of Architecture and Construction Sciences IUACC, Universidad de Sevilla, 41012 Seville, Spain

[3] Andalusian Housing and Rehabilitation Agency AVRA, Ministry of Development, Infrastructures and Spatial Planning, Junta de Andalucía, 41018 Sevilla, Spain; marial.perez.lopez@juntadeandalucia.es

* Correspondence: agarcia6@us.es; Tel.: +34-639-46-56-69

Abstract: The management of the life cycle of large publicly owned social housing complexes requires a large amount of human and technological resources, the optimization of which is a desirable and shared objective. This article proposes a workflow for the enhancement of these management processes based on BIM (Building Information Modelling), a methodology capable of integrating architectural information into a three-dimensional graphic model. The proposed workflow defines the basic characteristics of the BIM model oriented toward sustainable building management and its relationship with the key moments of its life cycle. It also analyzes the architectural information associated with the models and determines which parameters are optimal for their completion from the BIM models in terms of reliability, auditability, and automation. For this purpose, a case study has been developed for a multifamily residential building in Malaga (Spain), owned by the Andalusian Housing and Rehabilitation Agency AVRA, a public agency that manages a housing stock of more than 70,000 dwellings.

Keywords: social housing management; building information modelling (BIM); computerised maintenance management system (CMMS)

Citation: Castellano-Román, M.; Garcia-Martinez, A.; Pérez López, M.L. Social Housing Life Cycle Management: Workflow for the Enhancement of Digital Management Based on Building Information Modelling (BIM). *Sustainability* **2022**, *14*, 7488. https://doi.org/10.3390/su14127488

Academic Editor: Shervin Hashemi

Received: 2 January 2022
Accepted: 16 June 2022
Published: 20 June 2022

1. Introduction

The economic, social, and environmental sustainability of public and private buildings depends, to a large extent, on efficient and effective maintenance and conservation management during their useful life cycle. To this end, numerous IT tools have been designed in recent decades, from simple databases to specialized software known as Computerized Maintenance Management Systems (CMMS) [1]. These tools provide information on the key points of buildings managed by an organization, supporting the making of informed decisions regarding them [2].

To complement these digital resources for management, some authors have proposed the convenience of incorporating Building Information Modelling (BIM), making use of its capacity to integrate architectural information into a three-dimensional graphic model [3–5]. The advantages that stand out are the three-dimensional visual presentation of the model, its ability to generate effective communication between the agents involved (e.g., technicians, users, managers, and politicians), assistance in the control of maintenance (e.g., refurbishments, repair operations, and its costs,) and its usefulness as a bridge to other technological applications [6–15]. In parallel to the development of BIM, the digital maintenance and conservation phase has experienced significant progress [16–24].

In the specific context of the application of BIM to the management of social housing, references are scarcer. In Europe, a case study has been carried out in Triolo, France, which proposes a social housing management strategy based on the generation of a BIM model and the direct management of it [25]. It characterises the advantages of using BIM but has the limitations of lacking external management software and not addressing the issue of managing many buildings.

Another European study, located in the United Kingdom, provides experience of the use of BIM in the refurbishment process of a social housing building. This experience focuses on the time of the intervention and not so much on the subsequent management, highlighting the advantages of the fourth dimension of BIM, i.e., the time sequence in the models [26].

In Oceania, a New Zealand experience promoted by a public institution, the Wellington City Council, is of interest. This experience focuses on the benefits that BIM methodology can offer a public manager in predicting maintenance costs and planning management to extend the useful life of buildings. BIM models act as a data collector whose further management is in the process of development [27].

In the Americas, an experience developed in Chile has been published in which BIM is introduced to manage the moment when newly constructed social housing buildings are occupied by their inhabitants. It shares with the experience presented in this article attention to the phase of the useful life of the building but focuses on the characterization of damages in the building, recorded in the model itself and not in their management from external applications [28]. Finally, an experience from Brazil provides the perspective of client requirements in social housing projects using BIM, focusing on the need to adjust the information structure of the models to the objectives of the latter [29].

1.1. Problem Statement

Despite these advances, organizations that have to manage a large number of social housing estates find it very difficult to apply these technologies. The main reason for this is that the buildings they manage are, for the most part, buildings of a certain age, built before the widespread use of digital graphics and information tools. First, that implies major problems with alphanumeric document management, generally solved with basic database tools, and, to an even greater extent, problems of graphic information management, which is stored on paper, raster images from the digitization of paper plans, or, in the best of cases, on vector CAD support. Second, this alphanumeric and graphical information must be converted into an effective real estate decision-making assistant. To do so, the data must be reliable and auditable. Furthermore, if data correction and updating is necessary, the process should be as automated as possible.

This is the case for the Andalusian Housing and Rehabilitation Agency (AVRA), one of the public organizations with the largest number of dwellings managed, with 73,989 houses that provide accommodation to more than 300,000 people, on whose experience the case study will be based [30].

AVRA has wide and varied real state under its supervision, the management of which involves a major documentation problem, including graphics, which is often present in preventive conservation projects [31]. Generally, this documentation is on paper, the result of the predigital tradition of managing records through copies and reproducible means, typical of the last century. On other occasions, especially in the 1990s, paper deliveries were accompanied by copies on CD media of planimetry developed in CAD in a digital environment. The latter generally contain dwg and dxf files, whereby the information is systematized in folders, layers, and linked files, from which a printed dump can be obtained on paper or in PDF format. Despite this, dossiers have continued to appear in print well into the current century, either on paper or in PDF format, a format that was officially launched as an open standard in 2008 and published by the International Organization for Standardization (ISO) as ISO 32000-1.

In the General Archive of AVRA, this agency has only commissioned paper projects since its foundation in 2013. However, the public housing stock is made up of social housing that comes from various entities (ministries, councils, and municipal companies), most of which is from the second half of the last century, which are graphically recorded on paper and deposited in the Provincial Offices, not in the General Archive. Project copies on CD began to be ordered at the beginning of this century but have not been inventoried. This last medium usually presents the problem of difficult long-term conservation since it is highly vulnerable and contains information that is difficult to retrieve. In the case of the rehabilitation of historical buildings, or those built prior to the middle of the twentieth century, which represent a minority in relation to the total computation of the agency's social housing, there is a definite lack of updated documentation, although on rare occasions documentation is available in provincial or municipal historical archives.

Currently, AVRA uses software for the management of its social housing, named herein as CMMS-AVRA. It organizes a database that contains essential information on each managed building. The content and structure of this database are established in the Law 8/2013, of 26 June, on urban rehabilitation, regeneration, and renovation (BOE, 27 June 2013). This law describes the information required in a Building Assessment Report, a document standardized by the Spanish Government [32] that records the basic architectural characteristics and state of conservation of buildings. Based on this information, AVRA plans the rehabilitation actions to be carried out in each campaign.

1.2. Objectives

The main objective of this article is to show how the management of the digital lifecycle of public social housing parks can be improved with the use of the BIM methodology. To achieve this main objective, the following specific objectives are designed. The first is the proposal of a strategy to be managed for the graphic documentation of the buildings, capable of providing the necessary information to set up the BIM model.

The second specific objective is the analysis of the information structure of the building managers' own applications, in this case the CMSS-AVRA software, to select the key parameters whose determination can be improved from the BIM model. The aim is to select those that the BIM model can complete with guarantees of reliability, auditability, and automation.

The third specific objective is the characterization of the BIM model oriented to the sustainable management of its life cycle. This characterization will be determined in terms of a modelling strategy and linkage with the different situations in which the building may find itself during its life cycle.

The fourth specific objective is to resolve the connection of the BIM model with the building managers' own applications, in this case CMSS-AVRA software, enabling the flow of information between the two.

2. Methods

2.1. Current Workflow Analysis

Information on the current situation of the workflows carried out by AVRA in the management of its social housing was extracted from the performance of structured face-to-face and online interviews. These interviews were carried out by the research team in the period January-February 2021. The interviewees were the technicians responsible for the maintenance of the AVRA buildings. These interviews focused on the current common process that the agency is carrying out in the management of its social housing and the processes that could be optimized using the BIM methodology.

The information required for the analysis and evaluation of the improvement possibilities was obtained from unstructured interviews with various building managers (IT experts, architects, engineers, and managers). In these interviews, an attempt was made to determine the real needs of AVRA in relation to maintenance and the real possibilities of implementing BIM-based conservation management, considering the knowledge of these resources held by the technical personnel in charge of building conservation management.

The final evaluation of this process is presented in Section 1.1 as a problem statement. Although a management software is in place, CMSS-AVRA, the management model is highly analogue. The process of data entry and maintenance of the management system is essentially manual, and updated graphic documentation is rarely available. By not benefitting from the automation provided by methodologies such as BIM, maintenance is costly and inefficient, according to the managers themselves.

2.2. Case Study Selection

To achieve the objectives, one of the multifamily residential buildings managed by AVRA was taken as a case study (Figure 1). This building, built in Malaga, is representative of the type of property that AVRA manages, and lies both in the field of BIM management of social housing and in the generation of information models of existing buildings.

Figure 1. Case study, building on Lemus Street nº5 (Malaga). General Directorate of Cadaster and authors.

This building was constructed in 1987 and is located between Lemus Street and Plaza Bravo, in the Trinidad/Perchel neighborhood of Malaga. Its floor plan is rectangular, with two-storey pendent blocks connected through the interior patio and the first-floor gallery, evoking the image of an old neighborhood corral that previously existed at that location. It has two façades, Lemus Street and Plaza Bravo, from which the building is accessed. All the social housing in the promotion remains under a social rental scheme.

2.3. Digital Capture

The generation of the BIM model is based on the 2D CAD survey of the building, supplied by AVRA. In order to verify the degree of approximation of the model with the physical reality of the property and to evaluate the margin of error that the utilization of previous standard documentation supposes, two data captures were also made using digital technology: (a) a static laser scanner, based on taking data through static scanning locations; and (b) a hand-held scanner that performs data collection dynamically, accompanying the movement of a person moving through the property [33]. The equipment used was, firstly, a LEICA BLK360 IMAGING LASER SCANNER, capable of capturing 360,000 points per second, with a range of 60 m and a precision of 4 mm; it also includes thermal, laser, and visible light images, and is able to create a 360° scan in just 3 min. Secondly, a ZEB-REVO hand-held scanner was used, with a maximum range of 30 m, 43,000 points per second, and a relative precision of 1 to 3 cm.

The work carried out with the LEICA BLK360 IMAGING LASER SCANNER consisted of collecting data through 21 scanning locations in the common spaces of the building, starting from the centre of the courtyard. The interior of the dwellings was avoided since it would have required a scanning location for each room and would therefore have considerably increased the time of field work.

Through the LEICA Data Manager program, the download of the 21 BLK360 files took 60 min and their importing through the LEICA Cyclone Register program took another 60 min, while the recording of the 21 positions for their unification and adjustment took

120 min, and their exporting in rcp format, occupying 18 Gb of information, took another 210 min. A total of 555 min was spent, during which a massive data collection of points was obtained, including the management of the chromatic information of each point. This allowed dynamic images of optimal resolution to be obtained that provide a virtual tour of the building.

The metric capture using the ZEB-REVO hand scanner was carried out in two successive data collections. The first collection included the interior of one of the houses, the entire gallery, stairs, and a roof terrace. In the second data collection, the exterior façades were scanned (Figure 2). The starting point for each data collection was the same, with a duration of each moving image of 15 min.

Figure 2. Raw point cloud obtained from the scan: sectioned axonometry of the point cloud.

The clerical work consisted of dumping the point clouds from the scanners into universal format files, with an e57 extension from the scanner rental company, and of sending them by email link, which took approximately 10 min. Our team then imported these files into Autodesk Recap, to convert them into an Autodesk rcp file, compatible with Autodesk AutoCAD and Revit programs (Figure 3), which took another 10 min.

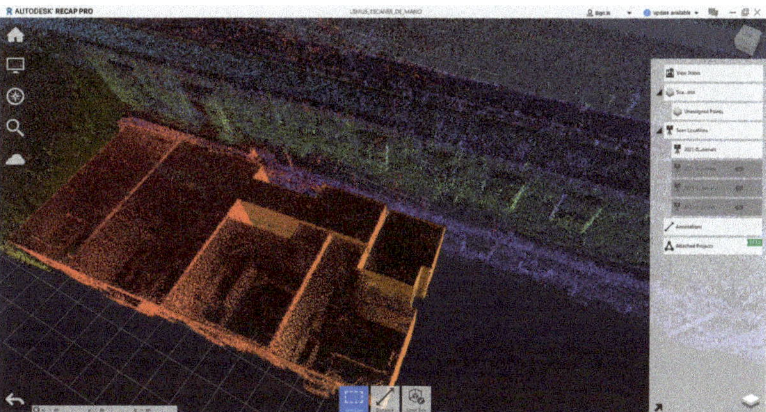

Figure 3. Point cloud after importing into Autodesk Recap.

This point cloud is scaled and oriented vertically from the origin, hence it is only necessary to orient it horizontally and move it based on a known georeferenced point. In our case, this cloud was transferred to the model that had already been prepared from the existing planimetry to verify the possible readjustments to be applied from this digital capture (Figure 4); this took another 10 min. The entire data collection and information management process to make it available in CAD was 60 min.

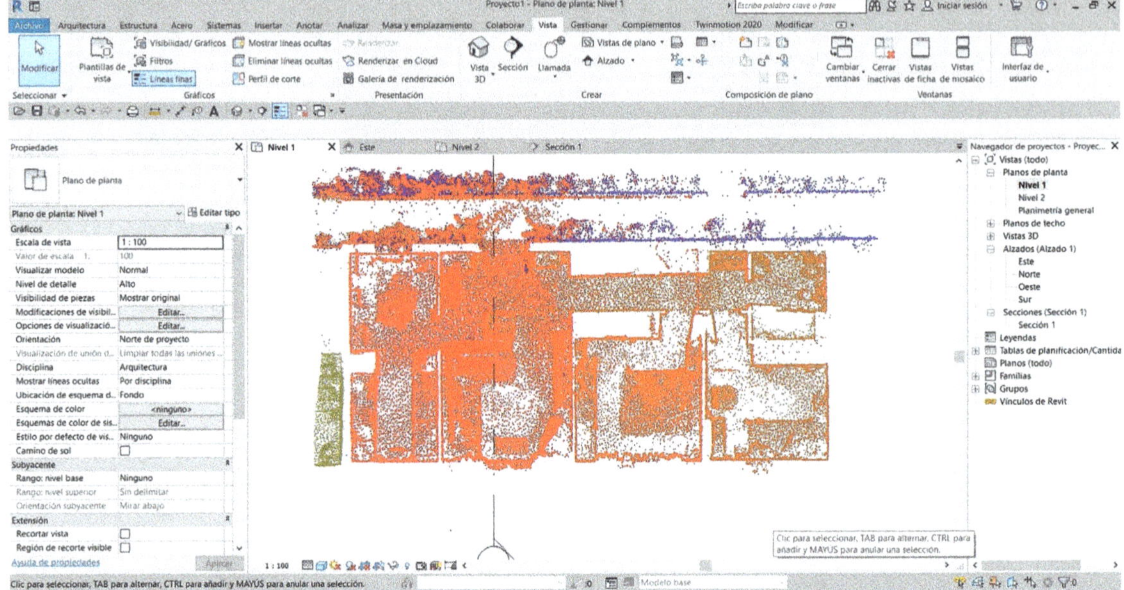

Figure 4. Model in Autodesk Revit with the imported point cloud.

A structured methodological process was designed according to the following sequence: analysis of the content and structure of the information of the AVRA program, development of the BIM model, and design of the input and output algorithms between the BIM model and the AVRA program.

2.4. Analysis of the Content and Structure of the Information in the Management Software

Information from CMSS-AVRA was received as data in Excel format. The large number of information fields resulted in the output of four separate files, three of which contained general property data and a fourth specifically oriented to the construction injuries registered in the building. The structure of this information was analyzed to identify which parameters allow a link between the CMSS-AVRA and the BIM model and which of these allow its automation from the said model.

2.5. BIM Model Generation

The BIM model was developed with the Autodesk Revit program. For the foundation of the model, CAD planimetric documentation and two three-dimensional scans were employed. However, the generation of the BIM model was not based exclusively on the metric information that could be extracted from the CAD vector graphics or from the point clouds of the three-dimensional scan, but rather required a constructive interpretation of the building. This interpretation was based on information obtained from the CMSS-AVRA and from direct inspection of the building itself (Figure 5).

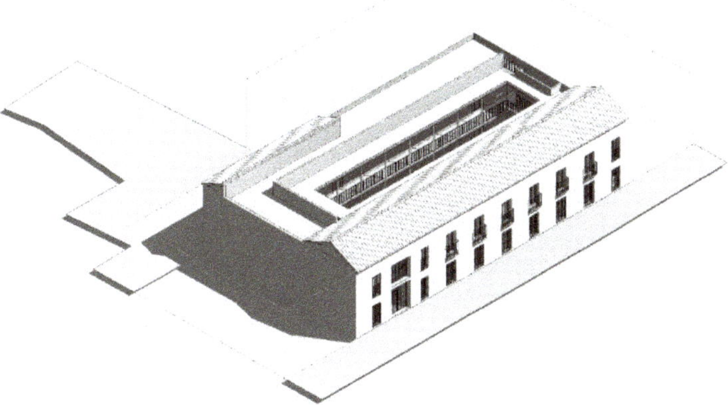

Figure 5. General 3D view of the BIM model.

2.6. Connection between Management Software and BIM Model

Once the BIM model had been formed and the information fields to be linked between the AVRA application and the model defined, the algorithms that make the transfer between them possible were then programmed. To this end, the Dynamo application, integrated in Revit, was used. An input algorithm was programmed from Excel, obtained from the AVRA application, and another output algorithm from the BIM Revit model to Excel: a format from which the AVRA application could input the information.

3. Results

3.1. Graphical Documentation Strategy

The implementation of the BIM methodology for the management of large property assets of institutions such as AVRA requires suitable systems and technological resources to render processes viable and sustainable over time. One of the first problems when building these models is the lack of a proper initial graphic documentation since most of these buildings were designed in a predigital era.

The starting point for the development of a BIM model lies in knowledge of the geometry of the building [34–36]. This can be obtained from a metric capture made expressly for the purpose or through an existing previous graphic. Nevertheless, on many occasions this documentation may not exist, may not be found, or does not accurately reflect the current state of the property. Four starting points were established:

a. There is an original prior planimetry, associated with administrative files, that is, hand drawings that were common between the end of the 19th century and the first half of the 20th century, usually in the form of floor plans and elevations, and occasionally as cross sections.
b. There is a description of the property as part of the architectural project or of the final state of the work, reproduced on paper, whose origin is an analogue drawing, common between 1950 and 1990.
c. There is previous digital vector documentation, generally in CAD, that has subsequently been put on paper for the management of the archive, generally from the 1990s onwards.
d. There is no graphic documentation, as they are very old files, or that it is a rehabilitation of a historical building (generally, up to the 19th century), which seldom retains systematic planimetry.

Given these situations, the development of a BIM model raises the choice between a metric capture made expressly for the purpose or the use of existing previous graphic documents. This is an important decision, since it can lead to an investment of considerable time and cost. The decisive factor is the purpose of the BIM model. If the building shows deformations and serious structural problems, or a major reformation is to be carried out thereon, then a comprehensive survey of its geometry would be necessary to develop a rigorous model in which to strictly quantify aspects that cannot be evaluated at first glance, such as overhangs. With the same starting point, if merely the daily preventive maintenance and registration of a building under normal conditions are envisaged, then not only can the model be much simpler, but the starting documentation can also be less metrically demanding. On the other hand, on many occasions, it is necessary to model the building in a simplified and sometimes repeated way for the development of computing applications for calculation, for example, for thermal efficiency.

Therefore, it is necessary to consider both questions, for which we start from Table 1, where the initial graphic information for the construction of the model is proposed according to its purpose.

Table 1. Relationship between geometric information support and model requirements.

		Geometric Information Needed			
		Previous Planimetry (a)	**Basic Data Collection (b)**	**Comprehensive Data Collection (c)**	**Not Necessary**
Objectives of the model	Record				✓
	Day-to-day management/ Preventive Conservation	✓			
	Specific Reforms	✓	✓		
	Energy Rehabilitation	✓	✓		
	Structural Rehabilitation			✓	
	Integral Rehabilitation			✓	

(a) Analog or digital; (b) sketch, measurement with conventional techniques; (c) digital capture/laser scanner.

For this research, both options were developed. On the one hand, the existing 2D CAD survey has been used for the generation of the BIM model. On the other hand, an alternative digital capture has been performed using both a manual scanner and a static scanner.

The information obtained from the scan was converted to rcp format, thereby obtaining point clouds that were linked to the BIM model. The comparison between point clouds is one more layer of information that enables the deformations of the elaborated model to be visually verified, and, if necessary, to be modified to better approximate the actual state of the building. In this case, the deformations occurred on the floor plan, due to the adjustment in the execution of the construction to the actual site, but these minor adjustments only affected the minor width of the floor and were insufficiently important as to modify the model (Figure 6).

From this analysis it can be deduced that in those cases where there is not a 2D digital model (CAD type), the best option for the realization of the generic BIM model for the management of the property would be to have a quick data collection of the property through the handheld scanner, which we have verified to be fast and accurate enough in most cases to subsequently obtain a model that is geometrically adjusted to reality. This type of instrument also makes it possible to accompany routine records carried out by technicians to verify any incident, since the transport of the instruments is light and the information management processes are fast.

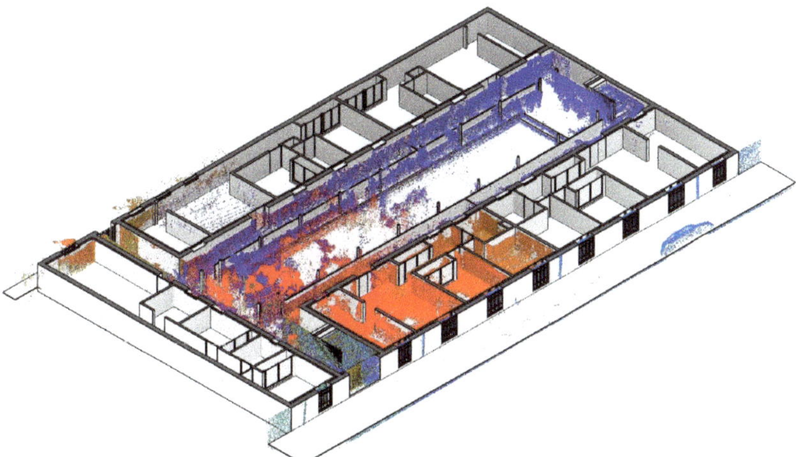

Figure 6. Cloud point and BIM model comparison.

Finally, the use of a ground scanner requires a greater investment in time and storage space, although its level of precision is much higher. Solutions to quickly record the data of various static captures already exist today, but typically produce a volume of information that can become unmanageable. For this reason, this type of metric capture is recommended for the analysis of structural problems, or the analysis of persistent pathologies, and should be avoided for problems that require a more generic definition of the model.

The use of the initial documentation requires, in turn, that it be located and linked to the database that contains not only all the records of the assets, but also the basic data regarding the surface area, location, etc. This provides a history to which future BIM models will be subsequently incorporated. The use of this entire graphic legacy has many other documentary functions that can be exploited in the long term, that concern the direct relationship between the forms of representation and the type of architecture developed [37].

3.2. *Information Structure Analysis*

As indicated previously, AVRA currently has software, CMMS-AVRA, for the management of its social housing. According to those responsible for this management, its effectiveness as a management program for the maintenance of its buildings has been sufficiently proven, at least in actions directed by the central services of the agency.

Consequently, the structure and content of the CMSS-AVRA database have been considered as the basis for the information structure to be included in the BIM model. For this, the information contained in the current CMMS-AVRA database was analyzed to determine how the use of BIM models could optimize the current action protocols. In the present investigation, the data related to the case study have been output from the CMSS-AVRA program in Excel format, since this facilitates the bidirectional transfer options with the BIM model. These data include the parameters of different formats and architectural significance, the relevance and usefulness of which in the model required their analysis and assessment. For this, two groups of data have been identified:

The first group includes the data that can be filled in indistinctly in the CMSS-AVRA or in the BIM model, that is, that in which none of the applications offers a significant advantage beyond the opportunity that some of them were already present. For example, the identification of the property will generally be completed in the AVRA program and could be output when a BIM model of one of them is generated. However, in the opposite direction, the foundation of a new BIM model on a property not yet registered in the AVRA application would allow the same identification data to be output (Table 2).

Table 2. Informative data that can be entered in both the CMSS-AVRA and the BIM model.

P_CODPRINEX	1099	**P_BARRIO**	Trinidad
P_MATRICULA	2070	**P_TIPOLOGIA**	P
P_DENOMINACION	MA-0994	**P_CALLE**	LEMUS
P_NUMVIVIENDAS	64-(MA-85/18-AS)/10 VPP	**P_CODPOSTAL**	29009
P_PROVINCIA	10	**P_REGIMEN**	ALQUILER
P_MUNICIPIO	MÁLAGA	**P_AÑOCONSTRUCC**	1987

The second group includes the data that can be extracted directly from the BIM model. These data provide the greatest potential for improvement among the social housing management processes. These data are related to architectural characterization (constructive, structural, functional, etc.) and its quantification. Being architectural information, these data are from the BIM model and, therefore, will have priority over those entered into the CMSS-AVRA. In general, they are quantitative data directly calculated by the BIM software from the modelled elements. For example, the constructed area of the property is an item of data that results from the BIM model and, therefore, is a graphically verifiable item of data that can be automatically updated if an error is detected. However, qualitative data derived from quantifiable data can also be defined. For example, the dominant joinery material is an item of qualitative data determined from the joinery surfaces of each material in the building (Table 3).

Table 3. Extract of quantitative data inherent to the BIM model, available to be exported to CMSS-AVRA.

E_SUP_PARCELA	542.5	**CV_OF_DA_SUPERFICIE**	722.15
E_SUP_CONSTRUIDA	972.75	**CV_OF_DA_PORCENTAJE**	72.45
E_ALTURA_RAS	8.10	**CV_CV_DA_SUPERFICIE**	106.82
CV_FP_DA_SUPERFICIE	167.75	**CV_CV_DA_PORCENTAJE**	10.72
CV_FP_DA_PORCENTAJE	16.83	**CV_AZ_DA_SUPERFICIE**	107.36
CV_FP_AR_PORCENTAJE	16.83	**CV_AZ_DA_PORCENTAJE**	29.6

From what is revealed in the analysis of the content and the information structure of the CMSS-AVRA, it can be inferred that the proposed BIM model must not be a mirror reflection thereof, but rather that the optimization of the information is proposed in terms of reliability, auditability, and automation: reliability in terms of the minimization of manual data transcription errors; auditability since the graphic condition of the BIM database allows both the visual recognition of the origin of the data in the three-dimensional model and the confirmation of its veracity; and automation in relation to the direct update of the data after the eventual modifications of the elements of the model. These specific advantages of employing BIM in this case study complement other advantages that are inherent in its nature as a graphical database and that are also useful in this context: the three-dimensional characterization and the availability of coherent planimetric documentation (Figure 7).

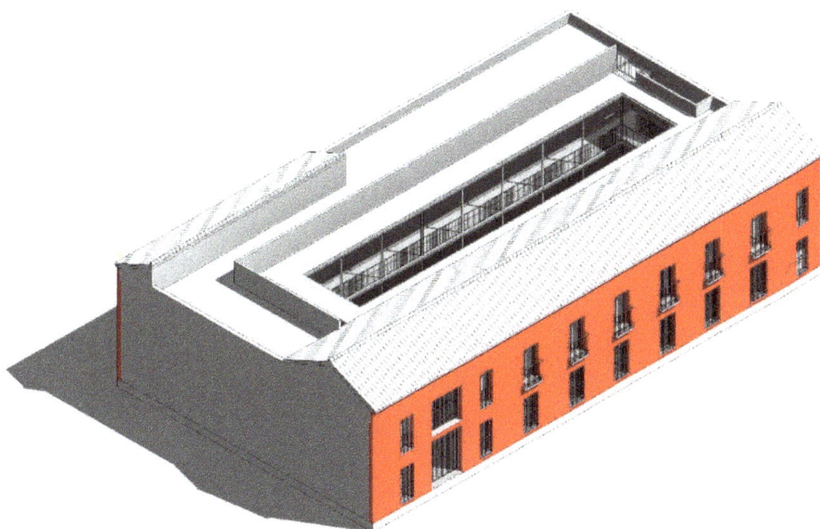

Figure 7. Data auditability. The parameter CV_FP_DA_SUPERFICIE filtered by color, which shows the area of the exterior façades.

3.3. *Generation and Characterization of the BIM Model*

The generation of the model involves both the geometric 3D modelling of the building and the design of the linked information structure, described in the previous sections. Geometric 3D modelling is based on a digital capture or on existing graphic documentation. In this case, both options have been explored to build the model, although the existing CAD survey has been sufficient to produce an operational model in accordance with AVRA's management requirements. Similarly, a linked information structure has been designed. The connection between the graphic elements and their information is solved thanks to the architectural analysis, which makes it possible to determine which information corresponds to each element of the model.

The resulting model is called BIM-AVRA, suitable for attending the management of a social housing manager. Its main characteristic is to be a 'live' model, i.e., a model that is systematically updated. To ensure the genuine effectiveness of this systematic update, the BIM-AVRA model must be as simple as necessary to ensure that the manager, in our case, AVRA, can bear the costs involved in the update process. In any case, the simplification must guarantee that the model includes the basic configuration of the building, but no elements that may be altered by third parties, whether tenants or other types of users. For example, the BIM-AVRA model will not include the color of an interior wall or the form of the bathroom cladding because these elements may be altered by third parties without the knowledge of the manager, thus rendering the model out of date (Figure 8).

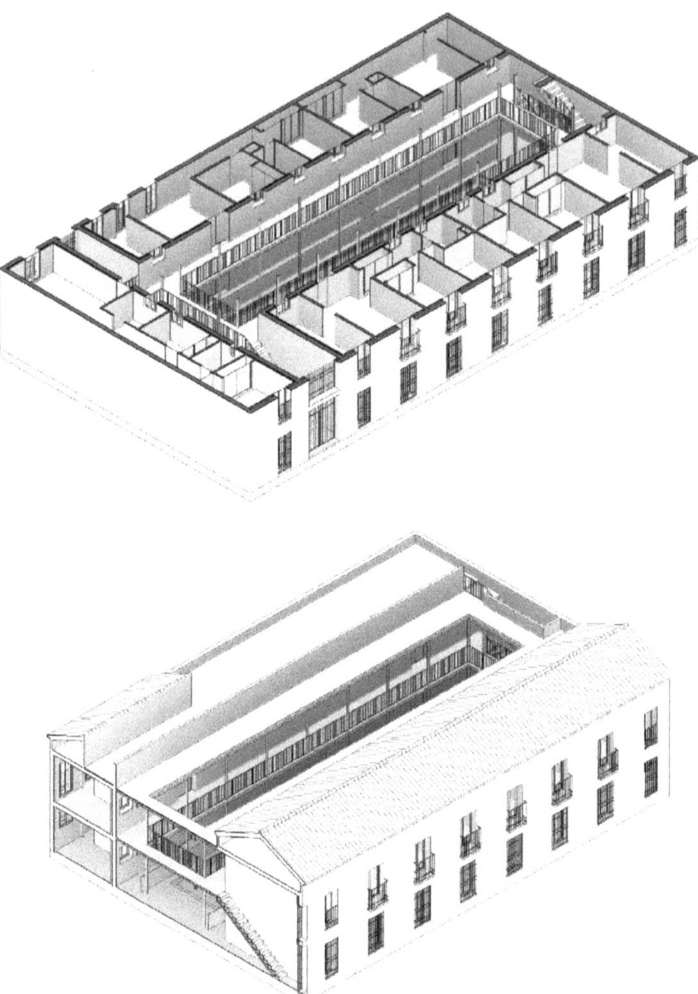

Figure 8. Horizontal and vertical section 3D views of the BIM-AVRA model of the case study.

Therefore, this model differs from others that could be created for specific new-build or refurbishment interventions. In other words, a single model cannot resolve the entire complexity of the useful life of a building, although it is possible to successfully integrate different models generated during its useful life. The BIM-AVRA model can offer basic architectural information about any of the agency's buildings and, therefore, can become the base model for any other models that may be generated for specific events during the useful life of a building, such as its extension, refurbishment, or demolition. These models would therefore serve a very different purpose, with the specific aim of documenting a design, execution, or state following an execution. Keeping these models separate from the BIM-AVRA model mitigates the problems derived from outsourcing the design and execution of a building because the models developed for these purposes will be independent from the base model and may be as detailed as required by the specific intervention addressed. Situations such as nonexecuted designs, partial studies, and building analyses, or even information regarded as an inaccurate reflection of the state of the existing building but useful to keep, will be archived as "still photographs" of the intervention without compromising the rigorous updating of the BIM-AVRA model. When these intervention models become substantial alterations of the building, they will be incorporated into the updated model in a simplified format consistent with the existing data in the model (Figure 9).

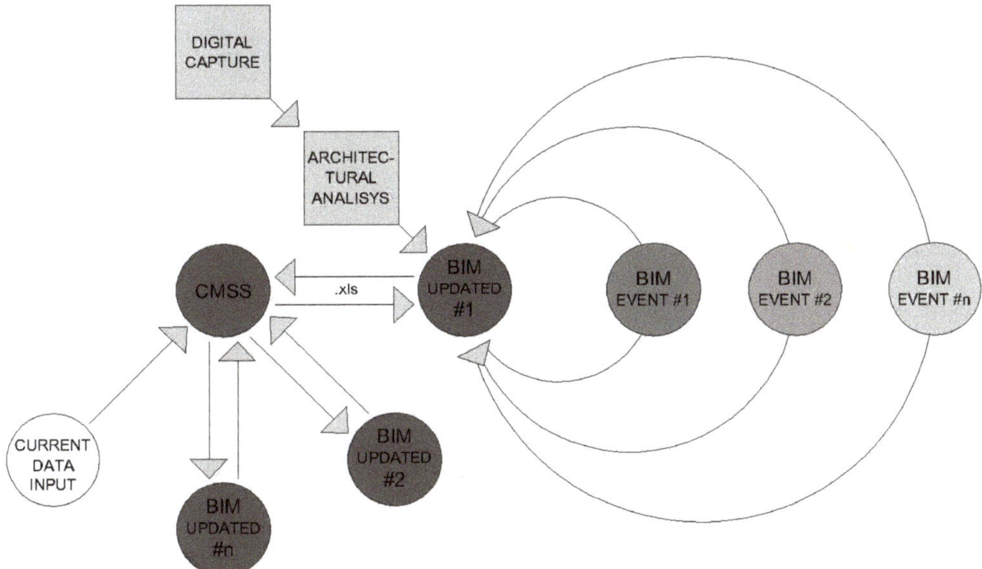

Figure 9. Diagram of the life cycle workflow around the updated BIM model.

3.4. Data Transfer and Interoperability BIM-AVRA/CMMS-AVRA

The algorithms that allow the transfer between the BIM model and the AVRA program are programmed with Dynamo, an application integrated in Revit that allows Excel files to be used as a format for the input and output of the model information. Both processes are structured by 'categories', that is, the set of objects that within a BIM model are associated with a certain constructive function (walls, roofs, doors, windows, etc.) or characterization of the model (information, project, built areas, etc.).

In the case of input from Excel, the Dynamo algorithm starts reading the selected data to assign them to the corresponding categories in the model. The process in the opposite direction is symmetric, although one must stop at articulating the data lists in a way that is consistent with the sequence expected by the AVRA management program (Figures 10 and 11).

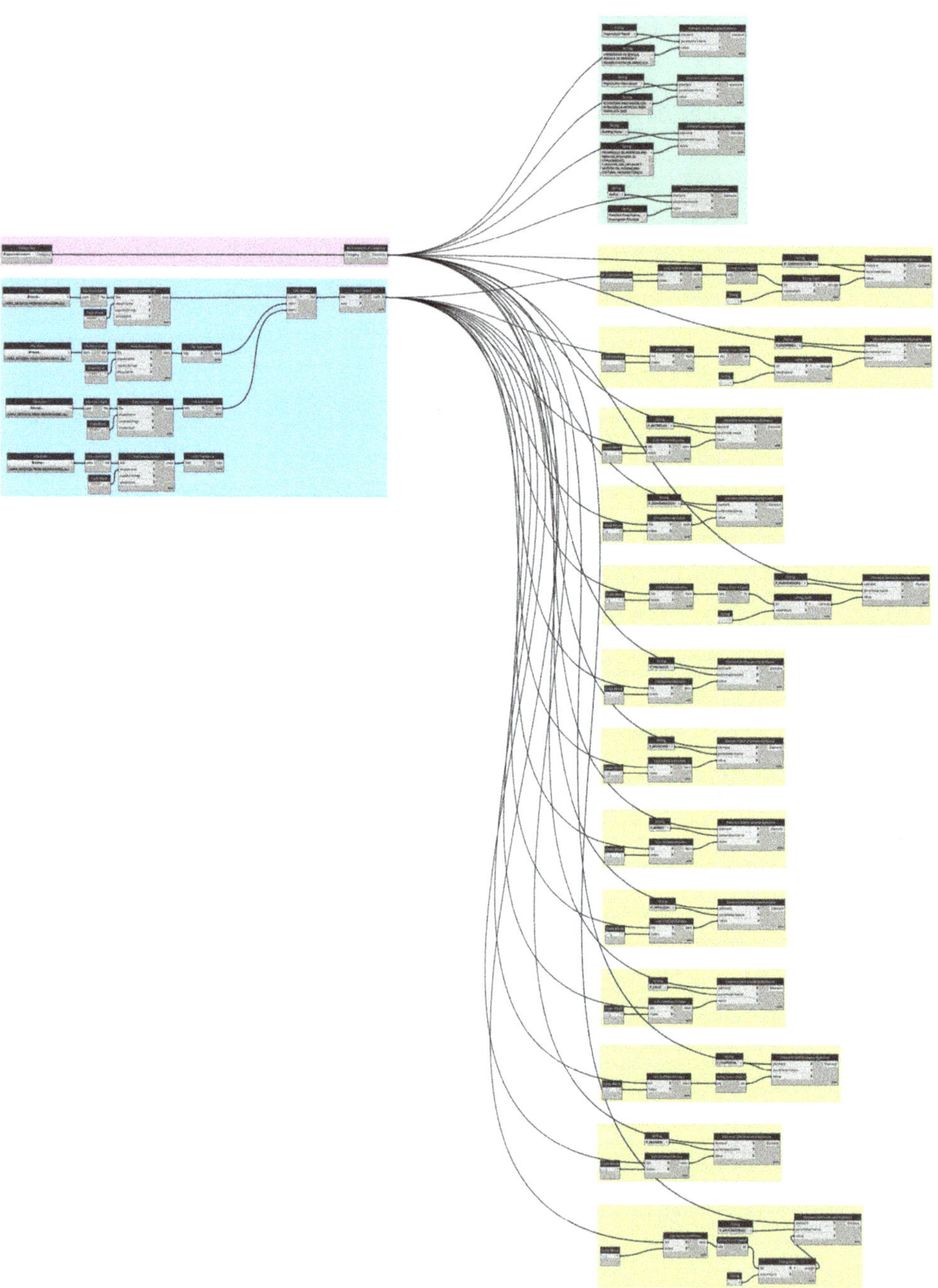

Figure 10. Dynamo algorithms for data output from Excel to the BIM-AVRA model (IMPORT.dyn).

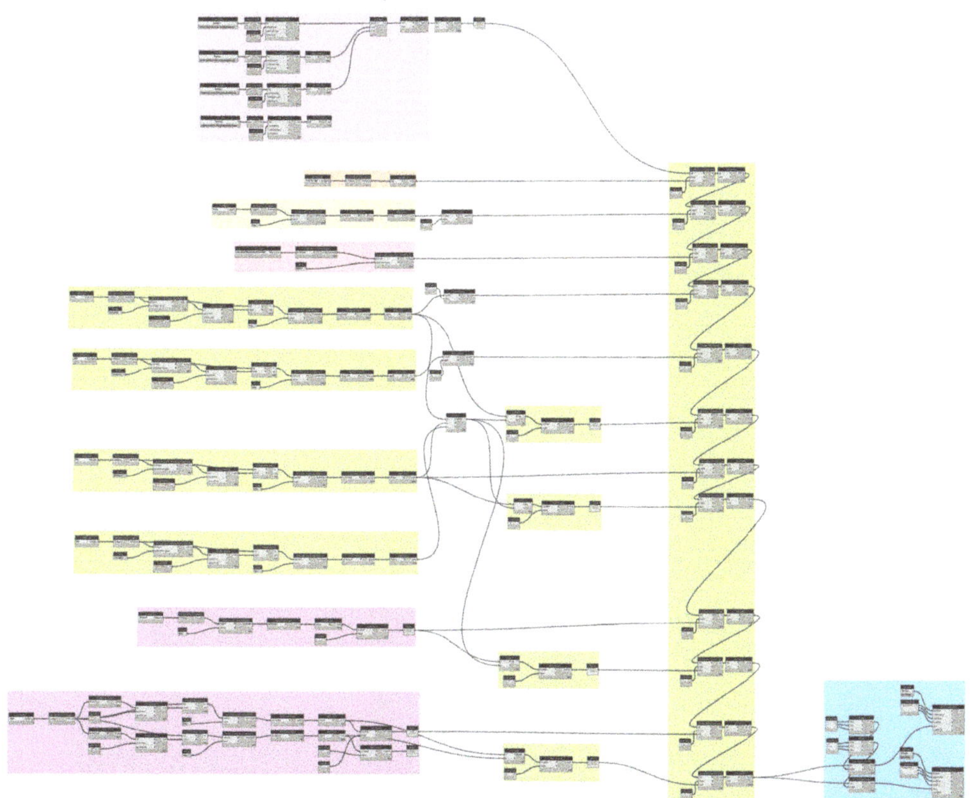

Figure 11. Dynamo algorithms for data output from the BIM-AVRA model to Excel (EXPORT.dyn).

3.5. Summary of the Proposed Workflow

The proposed workflow is organised around two main cores: the CMMS software of the managing institution, in our case CMMS-AVRA, and the updated BIM model of the building, in our case BIM-AVRA (Figure 9). The CMMS is a single core, and the BIM model is a multiple core, so there will be as many updated BIM models as there are buildings in the building stock to be managed. This does not mean that the workflow is only operational when all updated BIM models have been generated, but it can be combined with the current workflow. CMMS management can remain operational with the current data input system, which, as described in the Introduction, is mainly manual and, as updated BIM models of the buildings are generated, the system becomes more reliable, auditable, and automated.

Once the action on a building has been decided, the process of generating the updated BIM model begins. For this, existing graphic documentation or, where appropriate, digital capture will be used, in accordance with the strategies described in the previous points. Regardless of the sources of graphic documentation for the generation of the model, it is essential to perform an architectural analysis to basically characterise the structural system, the construction system, and the MEP system. This analysis allows the level of development of the model to be adapted to the knowledge available and to the resources available to the manager to keep it updated.

During the life cycle of the building, a number of unique events will occur, depending on the different types of intervention that the building manager may decide to carry out on the building. The updated BIM model provides reliable information for the intervention project and can serve as a basis for the generation of a BIM model specifically oriented to the objectives of that intervention. In other words, a BIM model of the corresponding event will be produced based on the updated BIM model, but different from it. Once the event is completed, the team responsible for maintaining the updated BIM model will revert the changes to the BIM model, according to its own standards and level of development. The BIM model used to produce the intervention will be archived and document the event at that point in the building life cycle.

At any time, the bidirectional communication of the updated BIM model with the CMMS can be updated via the xls export and import process described in the previous sections.

4. Discussion and Conclusions

This research presents a workflow to improve social housing life cycle management using BIM. To this end, the AVRA case study has been considered, given its role as a public owner that manages more than 75,000 dwellings. Based on unstructured interviews with those responsible for AVRA's property management, the following problems were stated: alphanumeric and graphic documentation and its use as an effective decision-making assistant for real estate.

The first step for BIM-based social housing management is the definition of a strategy for the graphic documentation of the buildings to be managed. The generation of the BIM model can be approached: (a) from 2D digitized plans (dwg, dxf or similar); (b) from plans on hardcopy; and/or (c) from the building itself by means of a digital and/or photogrammetric capture. Although digital capture offers advantages such as speed and accuracy, it is also a high-cost investment when it comes to hundreds of surveys. In the case study, two options were tested: the generation of the model from 2D CAD plans and a digital capture. It was concluded that the key point in making a choice is the purpose of the model, and a table was presented with the different purposes and recommended resources. For the day-to-day management of such a number of buildings as AVRA manages, the generation of the BIM model from 2D CAD drawings was proven to meet the proposed objectives.

On the other hand, digital capture by scanner was shown to be an effective and efficient procedure for generating BIM models. Two alternatives were studied for the metric capture of buildings: I. using a manual scanner; and II. a static scanner. It has been verified that, whatever the case may be, the information taken is compatible and can be related to the final BIM model, whether it is formed based on the digital capture carried out or not. The use of a manual scanner presents the advantages of transportable equipment, speed, simplicity, and efficiency of the data capture procedure compared to the static scanner, which requires more time for preparation and data collection. However, the static scanner presents greater precision. Given these results, it can be concluded that the point cloud obtained with the handheld scanner offers suitable precision for the realization of a general building management model. The point cloud obtained by means of the static scanner, more precise than the previous scanner, is useful in buildings that require intervention from a constructive and/or structural point of view.

The key parameters of the BIM model were determined to enhance the management carried out together with the CMSS-AVRA. This means that the information structure of the model is not a mirror of the CMSS-AVRA information, but the definition of two groups of data: those that are simply informative and those that quantify the architectural characteristics of the building. Informative data can be added to both software, but the source of quantitative architectural data must be the BIM model, as this ensures reliability, auditability, and automation.

The characterization of the BIM model oriented towards the sustainable management of the building's life cycle has been determined. In the modelling strategy, the requirement for simplicity prevails, modulated by the manager's capacity to carry out up-to-date maintenance that guarantees its long-term operability. Although simplified, the model must incorporate the basic configuration of the building, leaving out everything that can be altered by third parties. Consequently, it is ruled out that a single model can assume the full complexity of the building's life cycle, although it is possible to effectively integrate different models generated throughout the building's life cycle. A workflow in which interaction with the CMSS-AVRA is always carried out from this updated BIM-AVRA model is proposed. This updated BIM-AVRA model is offered as a basis for the development of other specific models for different events in the life of the building. In turn, the BIM models for each event will be archived and only the information relevant for the day-to-day maintenance of the building will be transferred to the BIM-AVRA model.

The connection between CMSS-AVRA and the BIM model has been solved, allowing the flow of information between both. A Dynamo script was programmed to automatically transfer the selected CMSS-AVRA data to the BIM model (IMPORT.dyn) via an Excel file. In turn, a Dynamo script was also programmed to automatically transfer the data contained in the BIM model to CMSS-AVRA via an Excel file (EXPORT.dyn). The latter, obtained directly from the BIM model, improves the management workflow, as they are reliable, auditable, and automated.

Finally, AVRA technicians have been shown the joint use of the CMSS-AVRA and the BIM-AVRA model of the case study, expressing their interest in the future developments of the results of this research. A first line of development would be oriented towards the extension of this experience to a sufficiently significant number of models to produce strategic decision-making assistance programmes. This would require selecting the key parameters of BIM models and combining them with others specifically designed to represent the strategic criteria of managers. A second line of development would be oriented towards the characterization of the modelling of singular buildings. In this sense, it would be particularly interesting to define specific strategies for the modelling of heritage architecture, incorporating artistic, historical, and archaeological information. Heritage applications of BIM, known as HBIM, have been widely tested in recent years, but it would be very interesting to reconcile their contributions with the management requirements of social housing use.

Author Contributions: Conceptualization, A.G.-M. and M.C.-R.; methodology, A.G.-M. and M.C.-R.; BIM model and Dynamo scripts, M.C.-R.; validation, M.C.-R., A.G.-M., M.L.P.L.; formal analysis, M.C.-R. and A.G.-M.; investigation, M.C.-R. and A.G.-M.; data curation, M.C.-R.; writing—original draft preparation, M.C.-R. and A.G.-M.; writing—review and editing, A.G.-M. and M.C.-R.; visualization, M.C.-R.; supervision, A.G.-M. All authors have read and agreed to the published version of the manuscript.

Funding: This research has been financed by European Regional Development Funds and the Junta de Andalucía, through the CEI-10-HUM799 project, within the framework of singular projects of transfer actions in the Campus of Excellence in the areas of the Research Strategy and Innovation (RIS3), within the global project of the University of Seville Innovative Ecosystem with Artificial Intelligence for Andalusia 2025.

Institutional Review Board Statement: Not applicable.

Informed Consent Statement: Not applicable.

Data Availability Statement: Not applicable.

Acknowledgments: The authors thank the Andalusian Housing and Rehabilitation Agency AVRA, the Ministry of Development, Infrastructures and Spatial Planning, and the Junta de Andalucía for providing all the information necessary for the development of the case study proposed in this investigation. We are especially grateful for the work carried out by Isabel Ramírez Acedo and Miguel Ángel Lobato Aguirre; the technical area members, Elena Morón, Mª María Bermejo Oroz, Jorge Ruiz García; the AVRA work team of the delegation of Málaga, Miguel Ángel Santos Amaya (Head of the Development and Quality Control Service, Ministry of Development, Infrastructure and Land Management—SGT); and Francisco Javier Ramos from the company Geoavance SL for his availability and kindness in accompanying and supporting the data collection. We also appreciate the contribution of the research project "Development of BIM-HBIM-GIS models applied to the knowledge, conservation, dissemination and management of architectural cultural heritage", especially the kind support of its coordinator Francisco Pinto Puerto and the researcher José María Guerrero Vega.

Conflicts of Interest: The authors declare no conflict of interest. The funders had no role in the design of the study; in the collection, analyses, or interpretation of data; in the writing of the manuscript, or in the decision to publish the results.

References

1. Cato, W.W.; Mobley, R.K. *Computer-Managed Maintenance Systems: A Step-by-Step Guide to Effective Management of Maintenance, Labor, and Inventory*; Butterworth-Heinemann: Oxford, UK, 2002.
2. Dzulkifli, N.; Sarbini, N.N.; Ibrahim, I.S.; Abidin, N.I.; Yahaya, F.M.; Azizan, N.Z.N. Review on maintenance issues toward building maintenance management best practices. *J. Build. Eng.* **2021**, *44*, 102985. [CrossRef]
3. Pärn, E.A.; Edwards, D.J.; Sing, M.C.P. The building information modelling trajectory in facilities management: A review. *Autom. Constr.* **2017**, *75*, 45–55. [CrossRef]
4. Gao, X.; Pishdad-Bozorgi, P. BIM-enabled facilities operation and maintenance: A review. *Adv. Eng. Inform.* **2019**, *39*, 227–247. [CrossRef]
5. Wong, J.K.W.; Ge, J.; He, S.X. Digitisation in facilities management: A literature review and future research directions. *Autom. Constr.* **2018**, *92*, 312–326. [CrossRef]
6. Yalcinkaya, M.; Singh, V.; Nenonen, S.; Junnonen, J.-M. *Evaluating the Usability Aspects of Construction Operation Building Information Exchange (COBie) Standard*; Tampere University of Technology: Tampere, Finland, 2016.
7. Akcamete, A.; Akinci, B.; Garrett, J.H. Potential utilization of building information models for planning maintenance activities. In Proceedings of the International Conference on Computing in Civil and Building Engineering, Nottingham, UK, 30 June–2 July 2010; Volume 2010, pp. 151–157.
8. Arayici, Y.; Onyenobi, T.; Egbu, C. Building information modelling (BIM) for facilities management (FM): The MediaCity case study approach. *Int. J. 3D Inf. Modeling* **2012**, *1*, 55–73. [CrossRef]
9. Aziz, N.D.; Nawawi, A.H.; Ariff, N.R.M. Building information modelling (BIM) in facilities management: Opportunities to be considered by facility managers. *Procedia Soc. Behav. Sci.* **2016**, *234*, 353–362. [CrossRef]
10. Brilakis, I.; Lourakis, M.; Sacks, R.; Savarese, S.; Christodoulou, S.; Teizer, J.; Makhmalba, A. Toward automated generation of parametric BIMs based on hybrid video and laser scanning data. *Adv. Eng. Inform.* **2010**, *24*, 456–465. [CrossRef]
11. Eastman, C.M.; Eastman, C.; Teicholz, P.; Sacks, R.; Liston, K. *BIM Handbook: A Guide to Building Information Modeling for Owners, Managers, Designers, Engineers and Contractors*; John Wiley & Sons: Hoboken, NJ, USA, 2011.
12. Kensek, K. BIM guidelines inform facilities management databases: A case study over time. *Buildings* **2015**, *5*, 899–916. [CrossRef]
13. Lu, Q.; Lee, S. Image-based technologies for constructing as-is building information models for existing buildings. *J. Comput. Civ. Eng.* **2017**, *31*, 04017005. [CrossRef]
14. Smith, D.K.; Tardif, M. *Building Information Modeling: A Strategic Implementation Guide for Architects, Engineers, Constructors, and Real Estate Asset Managers*; John Wiley & Sons: Hoboken, NJ, USA, 2009.
15. Volk, R.; Stengel, J.; Schultmann, F. Corrigendum to Building Information Modeling (BIM) for existing buildings—Literature review and future needs. *Autom. Constr.* **2014**, *38*, 109–127. [CrossRef]
16. Gunay, H.B.; Shen, W.; Newsham, G. Data analytics to improve building performance: A critical review. *Autom. Constr.* **2019**, *97*, 96–109. [CrossRef]
17. Pishdad-Bozorgi, P.; Gao, X.; Eastman, C.; Self, A.P. Planning and developing facility management-enabled building information model (FM-enabled BIM). *Autom. Constr.* **2018**, *87*, 22–38. [CrossRef]
18. Khudhair, A.; Li, H.; Ren, G.; Liu, S. Towards future BIM technology innovations: A bibliometric analysis of the literature. *Appl. Sci.* **2021**, *11*, 1232. [CrossRef]
19. Abuimara, T.; Hobson, B.W.; Gunay, B.; O'Brien, W.; Kane, M. Current state and future challenges in building management: Practitioner interviews and a literature review. *J. Build. Eng.* **2021**, *41*, 102803. [CrossRef]
20. El Ammari, K.; Hammad, A. Remote interactive collaboration in facilities management using BIM-based mixed reality. *Autom. Constr.* **2019**, *107*, 102940. [CrossRef]
21. Fumagalli, L.; Macchi, M.; Rapaccini, M. Computerized Maintenance Management Systems in SMEs: A survey in Italy and some remarks for the implementation of Condition Based Maintenance. *IFAC Proc. Vol.* **2009**, *42*, 1615–1619. [CrossRef]
22. Lai, J.H.K.; Yik, F.W.H. Hotel engineering facilities: A case study of maintenance performance. *Int. J. Hosp. Manag.* **2012**, *31*, 229–235. [CrossRef]
23. Lee, S.; Akin, Ö. Augmented reality-based computational fieldwork support for equipment operations and maintenance. *Autom. Constr.* **2011**, *20*, 338–352. [CrossRef]
24. Motamedi, A.; Hammad, A.; Asen, Y. Knowledge-assisted BIM-based visual analytics for failure root cause detection in facilities management. *Autom. Constr.* **2014**, *43*, 73–83. [CrossRef]
25. Alileche, L.; Shahrour, I. Use of BIM for Social Housing Management. In Proceedings of the 17th International Conference on Computing in Civil and Building Engineering, Tampere, Finland, 5–7 June 2018; pp. 1–8.
26. Chaves, F.; Tzortzopoulos, P.; Formoso, C.; Shin-Iti, J. Using 4D BIM in the Retrofit Process of Social Housing. In Proceedings of the ZEMCH Conference (Zero Energy Mass Custom Home), Lecce, Italy, 21–25 September 2015; Available online: http://eprints.hud.ac.uk/id/eprint/25563/ (accessed on 21 April 2022).

27. Wellington City Council Bracken Road Flats. Applying BIM Retrospectively as a Data Collection Tool for Maintaining Social Housing. BIM in Action Case Study 1. Available online: https://www.building.govt.nz/assets/Uploads/projects-and-consents/building-information-modelling/nz-bim-case-study-1-bracken-road.pdf (accessed on 21 April 2022).
28. Gonzalez-Caceres, A.; Bobadilla, A.; Karlshøj, J. Implementing Post-Occupancy Evaluation in Social Housing Complemented with BIM: A Case Study in Chile. *Build. Environ.* **2019**, *158*, 260–280. Available online: https://www.sciencedirect.com/science/article/pii/S0360132319303373 (accessed on 21 April 2022). [CrossRef]
29. Baldauf, J.P.; Formoso, C.T.; Tzortzopoulos, P.; Miron, L.I.G. Soliman-Junior, Using Building Information Modelling to Manage Client Requirements in Social Housing Projects. *Sustainability* **2018**, *12*, 2804. [CrossRef]
30. AVRA Institutional Information. Available online: https://www.juntadeandalucia.es/avra/opencms/areas/Informacion_institucional/index.html (accessed on 1 April 2022).
31. Public Residential Stock Improvement and Maintenance Plan. Available online: https://www.juntadeandalucia.es/sites/default/files/2020-03/2018_PMM_PPResidencial_CAA.pdf (accessed on 13 September 2021).
32. Government of Spain. *Building Assessment Report*; Government of Spain: Madrid, Spain. Available online: https://iee.fomento.gob.es/ (accessed on 22 April 2022).
33. Sepasgozar, S.M.E.; Forsythe, P.; Shirowzhan, S. Evaluation of terrestrial and mobile scanner technologies for part-built information modeling. *J. Constr. Eng. Manag.* **2021**, *144*, 8110. [CrossRef]
34. Brumana, R.; Oreni, D.; Barazzetti, L.; Cuca, B.; Previtali, M.; Banfi, F. *Survey and Scan to BIM Model for the Knowledge of Built Heritage and the Management of Conservation Activities, Digital Transformation of the Design, Construction and Management Processes of the Built Environment*; Daniotti, B., Gianinetto, M., Della Torre, S., Eds.; Springer: Berlin/Heidelberg, Germany, 2020; pp. 391–400. [CrossRef]
35. Rodríguez-Moreno, C.; Reinoso-Gordo, J.F.; Rivas-López, E.; Gómez-Blanco, A.; Ariza-López, F.J.; Ariza-López, I. From point cloud to BIM: An integrated workflow for documentation, research and modelling of architectural heritage. *Surv. Rev.* **2018**, *50*, 212–231. [CrossRef]
36. Wang, C.; Cho, Y.K. Performance evaluation of automatically generated BIM from laser scanner data for sustainability analyses. *Procedia Eng.* **2015**, *118*, 918–925. [CrossRef]
37. Puerto, F.P.; Vega, J.M.G. Imagen y modelo en la investigación del patrimonio arquitectónico. *Virtual Archaeol. Rev.* **2015**, *4*, 135–139. [CrossRef]

MDPI

Article

BIM and Automation in Complex Building Assessment

Jan Růžička [1,2,*], Jakub Veselka [1,2], Zdeněk Rudovský [3,4,5], Stanislav Vitásek [5] and Petr Hájek [2]

1 University Centre for Energy Efficient Buildings (UCEEB), Czech Technical University in Prague, 27343 Bustehrad, Czech Republic; jakub.veselka@cvut.cz
2 Department of Building Structures, Faculty of Civil Engineering, Czech Technical University in Prague, 16000 Prague, Czech Republic; petr.hajek@fsv.cvut.cz
3 Department of Construction and Investment, Czech Technical University in Prague, 16000 Prague, Czech Republic; zdenek.rudovsky@cvut.cz
4 Rector's Office, Czech Technical University in Prague, 16000 Prague, Czech Republic
5 Department of Construction Management and Economics, Faculty of Civil Engineering, Czech Technical University in Prague, 16000 Prague, Czech Republic; stanislav.vitasek@fsv.cvut.cz
* Correspondence: jan.ruzicka@fsv.cvut.cz

Abstract: When using Building Information Modeling (BIM) for complex building design, optimizing the building quality in a design phase becomes an important part of integrated and advanced building design. The use of data from an information model in the design phase allows efficient assessment of different design strategies and structural variants and a higher quality of the final design. This paper aims to analyze and verify possible BIM data-driven workflows for Complex Building Quality Assessment (CBQA) and a suitable BIM data structure set up for automatic assessment and evaluation. For an efficient automation process in complex quality building assessment in the design phase, it is necessary first to understand the data structure of the Industry Foundation Classes (IFC), which is widely accepted and used for buildings, and second to understand the data structure of the assessment methodology used for the assessment. This article describes possible data workflows for an automatic assessment based on the experience gained on a case study of the real pilot project of a residential building, where the complex building quality was tested using SBToolCZ, the Czech national assessment method. This article presents the experience and recommendations for setting up the data model of a building for automatic assessment.

Keywords: Building Information Modeling (BIM); Life-Cycle Assessment (LCA); building optimization; Industry Founded Classes (IFC); SBToolCZ

Citation: Růžička, J.; Veselka, J.; Rudovský, Z.; Vitásek, S.; Hájek, P. BIM and Automation in Complex Building Assessment. *Sustainability* **2022**, *14*, 2237. https://doi.org/10.3390/su14042237

Academic Editor: Ali Bahadori-Jahromi

Received: 31 December 2021
Accepted: 11 February 2022
Published: 16 February 2022

1. Introduction

The complex quality of buildings is one of the main issues of the current building sector, which covers a wide range of criteria, including environmental quality and LCA, energy performance of buildings, durability, and resilience, quality of the internal environment, architectural and functional quality, and other social issues. Current rating systems and assessment methods and frameworks are efficient, inspiring, and motivating tools in the design process to achieve the goal of complex building quality. The complexity of the rating systems, on the one hand, and a huge number of criteria and parameters required for the assessment, on the other hand, are a challenge when searching for new ways for automation in this field. Connecting Building Information Modeling (BIM) as a structured database of building parameters with sufficient data workflows linked to the assessment scheme can provide an efficient tool for complex building quality assessment.

BIM is still a new idea in construction, representing significant technological progress in information technology that has been visible in recent years. It combines a digital environment with a real one and has great potential to change traditional ways of working in the design, construction, and management of construction projects. The BIM method has several definitions from leading construction experts and building associations. The most frequently inflected interpretation gives the authors of the article a definition: "BIM technology represents a digitally created one or more precise virtual building models. Supporting object design across all phases of the object life cycle, enabling better analysis and control than manual (traditional) processes. When these computer-generated models are ready, they contain the exact geometry and data needed for the construction, production, and procurement process according to which the building is carried out, including subsequent management" [1]. In general, all interpretations associated with the definition of the BIM method have common features based on the following:

- Sharing information between individual entities in the design, construction, and subsequent management of the building;
- The 3D visualization and localization of structures or the entire building;
- Faster decision-making associated with project management and overall higher economic efficiency of the project.

However, according to the perspective of the global consulting company McKinsey & Company and its Industry Digitization Index, the Architecture, Engineering, and Construction (AEC) field is still one of the least digitalized fields of human activities [2,3]. Other recognized publications express in a similar vein the views that the construction industry is among the least digitized industries, and the lack of innovation in construction project management practices has led to decreased productivity [4]. This includes the point of view of public contracting authorities (public sector), who have modern technologies required for more efficient spending of public funds, where selected articles recommend starting to "push" BIM in this way and thus start market development [5–7]. It also includes the point of view of private investors, where the "correct" grasp of these modern processes carries the private companies themselves into the new digital era, and therefore, there are desired savings of time and, especially, costs [8–10]. In the overview of professional publications and studies, the authors in the following subchapters focused on topics that are closely related to the subject of the article, where their conclusions and messages were considered in addressing the issue (in addition to obvious benefits and pitfalls of the digital implementation process).

1.1. Sustainability of Construction

The sustainability of construction, together with its digitization, is one of the most current topics, both in the field of research and in construction practice. Recent studies based on more than 650 buildings produced by several research groups [4,11–13] show that in the following decades, the importance of embodied greenhouse gas (GHG) emissions is greater than operational emissions. BIM can be involved in many particular tasks within design, construction, and operation within this issue. Relevant Asian projects linking BIM and sustainability were in Malaysia, China, and Kazakhstan. In all cases, it was a strategic decision to place a BIM method to determine the degree of sustainability in the solution of a project identified as sustainable, including the subsequent use of data for its management. The conclusions of the articles are related to balancing the advantages and disadvantages of digital technologies, where the benefits of these modern techniques clearly prevail [14–16]. Another important article on this issue is research on the Korean environment, which focuses on the impact of the BIM method and sustainability on high-rise buildings. The detection process took place through Exploratory Factor Analysis (EFA) and Structural Equation Modeling (SEM). The result of the research was the identification of Critical Success Factors (CSFs) for the implementation of BIM on sustainable and high-rise construction, where 205 construction projects participated in the questionnaire survey. Key criteria identified by this research were included in a modified form in the solution of their article [17].

The authors of practical studies have also developed them in the energy sector. Specifically, these are two articles focused on applying BIM in the field of energy with a direct link to sustainable topics. Both articles deal with identifying data needs and subsequent analyses for the energy and sustainable model of building in the European environment.

The primary goal was to automatically generate an accurate and flexible model called Building Energy Modeling (BEM). BEM made it possible to automate the calculations of primary energy savings and CO_2 emissions for the solved projects and thus met the pilot objectives of the project [18,19]. However, from the point of view of the authors of their own article, some visions were too progressive and may have a practical impact on routine operation in the next few years.

In general, the topic of building sustainability has seen significant progress in recent years, both in research and in practice, where several digital tools have been developed to allow calculations based on input parameters. It is these input parameters that BIM can provide if entered correctly. In creating their own article, the authors used the applied methods in selected publications, primarily in connection with the method of defining data needs for similarly large projects intended for sustainable construction.

1.2. Building Life Cycle

Following the current text, the BIM method has the greatest use throughout the life cycle. That is, from the construction design through its implementation and subsequent administration. There are a number of examples that confirm BIM as a potential life-cycle method [9,10]. Within the design of constructions, there are several dozen successful case studies, and not only from the European environment. However, interesting information seems to focus on the individual "less-feasible" steps in the project design. In particular, it is the implementation of systems based on the Common Data Environment (CDE), where all participants in the construction design must learn to work with and use modern tools for their routine activities. Alternatively, additional information associated with coordination activities between several construction objects is available if more than one design program is used and "meets" in one data environment (primarily a matter of line constructions) [20–23]. These findings are valuable for considering how to realize the implementation process of the BIM method, not only in the stage of construction design.

The next stage of the construction life cycle associated with monitoring the progress on a construction site during the construction phase is crucial. Construction monitoring using the BIM method enables digital tools primarily associated with construction costs and schedules. An interesting observation in this regard is the use of Field Data Capture Technologies (FDCT) and Communication and Collaborative Technologies (CT), where colleagues applied digitization in a real construction project from the Spanish environment. Other articles also present other modern technologies digitizing processes in construction, such as Dynamic Site Layout Planning (DSLP) and Development and Framework for Integrating Unmanned Aerial Vehicles (UAVs) [24–27]. The conclusions of these articles and studies have common denominators in the form of three pieces of information. The first is that setting up a project for these digital processes should ideally follow the previous design stage. Furthermore, "data discipline" is required, where the data in all systems must be constantly updated, and, last but not least, the key is the support of the construction management, which determines the direction of the construction process.

By far, the longest stage in the construction life cycle is its management. There are several professional publications dealing with the benefits of involving the BIM method in the operation of both buildings and line objects. Leading publications present these benefits in specific projects. In particular, they deal with the integration of data from BIM models into various Computer-Aided Facilities Management (CAFM) systems with an emphasis on the greatest possible automation of information transfer. In a real operation, this information has a key role in the maintenance and servicing of individual machines/structures, including their location, etc. [28–30]. The result of this "lifelong" data interconnection is a Digital Twin, which represents the target state of the digitization of the building environment.

1.3. Life-Cycle Assessment (LCA) and Complex Building Quality Assessment (CBQA)

Another important methodology recently highly used in the AEC sector is Life-Cycle Assessment (LCA). This method can evaluate the negative environmental impacts of the building and help optimize it. The LCA method is described in the ISO 14044:2006 norm [31] and can be adjusted according to the national standards, which can vary by 20% [32,33].

BIM and LCA can be combined in various use cases, as shown in recent articles [11,12,34,35]. The purpose of all studies is to develop a workflow that can be (i) precise enough for the building design process, (ii) possible for building optimization (in terms of structure, materials, environmental impacts, etc.), and (iii) replicable for another project with minimum effort. Wastiels [36] wrote an overview of possible different workflows [37], and Obrecht et al. processed a systematic literature review [13].

The whole LCA method is very complex and not perfectly suitable for everyday practice. It also only covers the environmental part of the building quality. Therefore, certification tools and methods for Complex Building Quality Assessment (CBQA), such as the Building Research Establishment Environmental Assessment Method (BREEAM), Leadership in Energy and Environmental Design (LEED), Deutsche Gesellschaft für Nachhaltiges Bauen (DGNB), and the SBTool were developed [38]. These tools simplify the LCA method, and moreover, they evaluate other aspects of the building quality: social, economic, and local. As a result, complex quality assessments can be processed [36,39]. In the reality of the Czech AEC market, the mentioned complex quality assessment methodologies are commonly used. In addition, the Czech national methodology SBToolCZ [40] was developed for the Czech market by the Czech Technical University in Prague together with The Technical and Test Institute for Construction Prague and the Building Research Institute—Certification Company. This assessment framework is based on the international SBTool and adjusted for national economic, social, and natural conditions. It is accessible for free for the following types of buildings: family houses, apartment houses, offices, and schools. The current target is to transfer the manual assessment process to the BIM-based workflow. The study's objective is to identify possible data workflows for combining BIM and SBToolCZ as an example of a complex quality rating system.

1.4. Industry Founded Classes

BIM can be used in any of the use cases. However, with higher demands of automation (e.g., analysis, simulations, model data mining, dashboards, e-building permit, etc.), the importance of standardization grows. At a certain level, it is already implemented in the mentioned norm ISO 16739-1:2018 [41]. The standardized data structure can be (i) included in the primary model developed in the BIM authoring tool or (ii) exported from the native format to the Industry Founded Classes (IFC). This open international format is developed by the Building Smart organization.

The IFC format represents a standard that the construction market must fully implement and adopt. Several publications deal with this standard and emphasize its purpose in achieving the most open construction process possible [42–44]. This activity is called OpenBIM. The conclusions of the publications also call for the continued development of this format, including related key support software. IFC can also be used during the building operation period [45].

As Obrecht et al. [13] state, there are many examples and case studies integrating BIM and LCA, and different workflows for combining BIM and LCA proposed by Wastiels [36] can be used. LCA represents a simplified assessment process because of the limited number of required parameters and set of properties compared to complex quality building assessment and its complexity. Integrating BIM and CBQA represents new challenges and raises new research questions. This article aims to share the experience gained with semi-automatic BIM data-driven workflows and a suitable BIM data structure set up for complex building quality assessment (CBQA) on a pilot case study of a residential building combining BIM and SBToolCZ.

2. Methodology

The main problem of an automatic complex quality building assessment is that current methodologies and assessment schemes do not follow the data structure of a Building Information Model. Due to a different data structure, it is impossible to use automatic algorithms.

The problem, on the one hand, is a completely different set-up of the assessment schemes, which use specific quality rating scales, often based on verbal or visual rating scales. The other issue is the Level Of Information (LOI) of the assessed Building Information Model (BIM). For an efficient automation process in complex quality assessment of the building in the design phase, it is necessary to first understand the data structure of the IFC, and second, to understand the data structure of the assessment methodology used for the specific assessment.

2.1. Data Format of a Building Information Model (BIM)

The Building Information Model (BIM) is a database of graphical and nongraphical data stored as (i) an open and vendor-neutral data file format or (ii) a proprietary data file format. Using an open and vendor-neutral data file format is crucial for public procurement projects, as well as for wide collaboration between different branches of the construction sector and between different stakeholders in the design and construction process. Thus far, the only widely accepted representative of data schemas in the construction industry is IFC (Industry Foundation Classes).

2.2. IFC Data Structure

IFC is a widely accepted, used, well-structured, standardized, open, and vendor-neutral data structure (or schema) of a built environment, including buildings and civil infrastructure. The IFC is maintained and developed by BuildingSMART [46], the non-profit organization, and it is an official standard ISO 16739-1:2018 [41]. The most important schemas of the IFC data format involve:

- Semantics and classification of data objects;
- Object properties;
- Object relationships.

The IFC data can be physically encoded in various formats. The most common is the STEP format (.ifc). An example of the data structure is shown in Figure 1.

2.2.1. Semantics and Classification of Data Objects

The object definition in IFC is anything perceivable or conceivable that has a distinct existence, albeit not material. It is a broad term. Therefore, there is a logical classification of all objects in IFC. All of the classified objects are rooted entities; they are rooted in a hierarchical classification tree. This classification integrates entities in a logical and characteristic inheritance manner. The subordinate (lower branch) entity inherits all characteristics of the superordinate (higher branch) entity and adds some of its own characteristics.

The detailed explanation of the IFC structure that is relevant for data modeling for complex quality building assessment is as follows:

- Superordinate entity: IfcObject is the generalization of any semantically treated thing or process.
    - Subordinate entity: IfcProduct: The IfcProduct is an abstract representation of any object that relates to a geometric or spatial context. The IfcProduct consists of annotations (cotes, comments, etc.), structural items (actions and reactions), spatial elements (site, building, story, space, spatial zone), Architectural, Engineering, and Construction elements (AEC products), etc.
        - Subordinate entity: IfcSpatialElement is the generalization of all spatial elements that might be used to define a spatial structure. It contains spatial zone (IfcSpatialZone), space (IfcSpace), story (IfcBuildingStorey), building (IfcBuilding), and site (IfcSite).

- Subordinate entity: IfcElement is a generalization of all components that make up an AEC product. It is IfcProduct with the addition of material nature.

IFC objects are specific modeled products in the building information model, such as columns, slabs, boilers, parking places, and zones for disabled users, which carry information about the properties, quantity, and quality that can be used for complex quality building assessment.

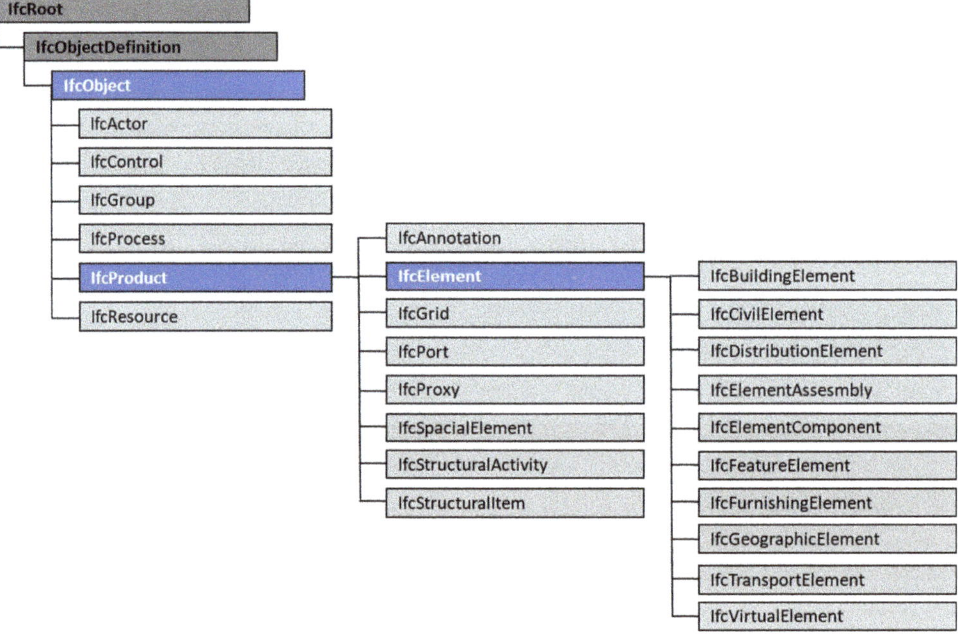

Figure 1. An example of the IFC structure [46].

2.2.2. Properties

Properties are, in general, described in IFC in two ways:

- A direct (or derived) object attribute (e.g., "Name" for each object);
- The individual property of a property set.

In addition, there is another structure of property value types mainly used:

- String, (e.g., "IfcText, IfcLabel, IfcInteger, IfcReal, IfcLenghtMeasure, IfcPowerMeasure, etc.")—text string or enumerations;
- Logical (e.g., "IfcBoolean, IfcLogical, etc.")—true or false (0/1).

Some alphanumerical values are enumerated.

As mentioned above, rooted entities are classified in the inheritance tree. This also applies to the inheritance of properties of rooted entities.

Example (Property set—property):

- Superordinate entity:

IfcElement: Has property set describing warranty (Pset_Warranty), that consists of 8 properties.

- Subordinate entity:

IfcActuator: Has a Pset_Warranty and, in addition, many other property sets: e.g., for electrical device common properties.

2.2.3. Relationship

Relationships among entities in IFC are realized by objectified relationships. There are two different types of relationships: one-to-one relationships, and one-to-many relationships. Objectified relationships express:

- Assignment;
- Association;
- Connectivity;
- Decomposition;
- Other relationships.

Objectified relationships are inverse attributes of the modeled object.

As mentioned above, rooted entities are classified in the inheritance tree. This also applies to the inheritance of rooted entity relationships. For example:

- Superordinate entity:

IfcSpatialStructureElement: The spatial decomposition is a logical structure of spatial elements ultimately assigned to the project. The order of spatial structure elements being included in the concept for building projects are, from high to low level, IfcProject, IfcSite, IfcBuilding, IfcBuildingStorey, and IfcSpace.

- Subordinate entity:

IfcSpace: Space is part of spatial decomposition and, in addition, has a relationship to covering elements. Those coverings may represent different flooring or tiling areas.

2.3. *BIM Data-Driven Workflows*

The aim of data mining from the information model is to analyze the building and optimize its design. There are several possibilities of BIM data-driven workflows depending on the level of digitalization in the building sector, on the development of national BIM data standards, and assessment methods. The goal is to achieve a fully automatic assessment workflow that allows real-time design optimization. The degree of automation highly depends on data standardization, on the level of integration of assessment methodologies in the BIM software, etc. The levels of model data structure and data-driven workflows are: (i) low structured model data, and manual workflow, (ii) utilized IFC data structure and semi-automatic workflow, and (iii) highly structured model data and automatic workflow.

2.3.1. Low Structured Model Data and Manual Workflow

In this case, model data are manually read, and parallel calculations and/or analysis are carried out in an independent calculation software (Figure 2). The results are manually inserted as model properties of a particularly added property set. The assessment information is not connected to modeled entities. BIM can exist without modeled entities, used just as a data format. This approach is inconvenient; design optimization is time-consuming and generates a high risk of calculating errors and errors in communication processes between the author of the building information model and authors of the specific calculations.

2.3.2. Utilized IFC Data Structure and Semi-Automatic Workflow

A semi-automatic workflow (Figure 3) uses calculations and analysis outside the data model in specific calculation software and tools, but the information needed for the assessment is exported in a predefined structure and can be automatically imported into the calculation software. Model data can be exported straight from the native model to the calculation tool in case the calculation software can read the data directly. This allows keeping the original data quality from the model. The model data from the native model can be exported in IFC format and then imported through the utilized database (e.g., .xls, .csv) or directly to the calculation software. Particular property or property sets and/or entities that are not common parts of the data model can be manually added in the calculation software. The results of the calculation and assessment can be uploaded back to the native model automatically in case of nongraphical information using sufficient data bridge or manually in case of graphical information. In the case of semi-automatic workflow, the time-consuming and manual part of the assessment process still remains, but communication processes and data exchange are efficient, and the risks of errors decrease.

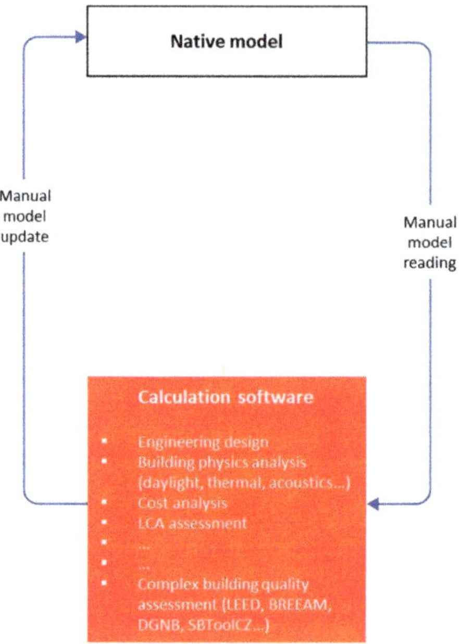

Figure 2. Low structured model data with manual workflow.

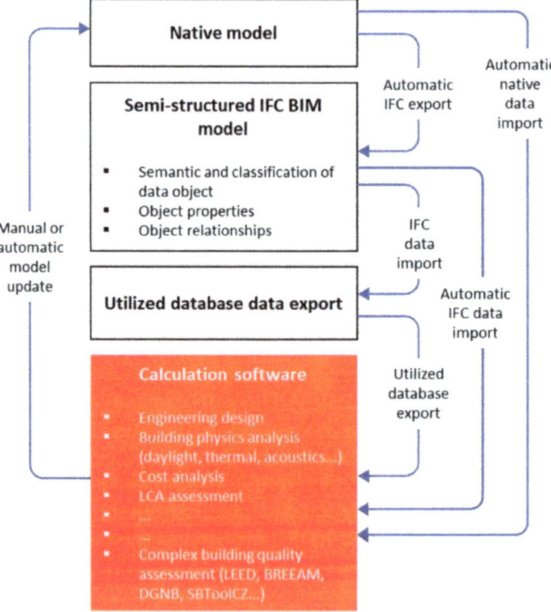

Figure 3. Utilized IFC data structure and semi-automatic workflow.

2.3.3. Highly Structured Model Data and Fully Automatic Workflow

The fully automated workflow (Figure 4) uses calculations and analysis as an integral part of the software for data modeling (as a plug-in calculation module) or as an independent calculation software and uses model data for calculations and analysis. External data input, if needed, is inserted automatically and directly to the model or as boundary conditions for the calculations. The calculation algorithms are prepared in the software (as an integral part or as a third-party plug-in), and they are calculated or analyzed in the model as well. As in the previous case, the results of the calculation and assessment can be uploaded back to the native model automatically in case of nongraphical information using sufficient data bridge or manually in case of graphical information.

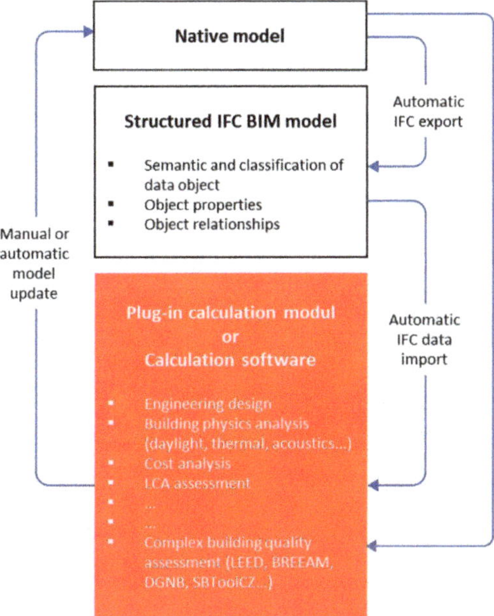

Figure 4. Highly structured model data and fully automatic workflow.

In this case, a high-level model data structure is needed. All modeled data are general and usable for other purposes, e.g., housing unit area quantities, housing unit quantities, etc. Assessment calculations are generated automatically. The main advantage is that in case of any design change, information take-off, calculation, and assessment are automatically generated in real-time with minimized risk of errors and misunderstandings, allowing efficient optimization of the designed building.

This approach requires powerful HW, on the one hand, and continues software maintenance and control according to changing standards, legislation, software version, etc.

2.4. Data Structure of SBToolCZ

2.4.1. SBToolCZ

To set up the architecture of the data model of the building, it is necessary to understand the data structure of the assessment methodology. This section describes the data structure of the Czech national assessment methodology SBToolCZ for sustainable and complex building quality assessment. The data structure of SBToolCZ is, in general, very similar to other assessment methods, such as LEED and BREEAM. Therefore, the approach of using data structures analysis can also be used for other assessment methods [37]. Data analysis was carried out on the assessment version for residential buildings.

SBToolCZ [40] is a national sustainability rating system developed for the Czech Republic that respects local building tradition, the social and economic environment, and its specifics. SBToolCZ is based on a generic framework SBTool (Sustainable Building Tool) for evaluating the sustainable performance of buildings, and projects developed by iiSBE (International Initiative for a Sustainable Built Environment) have been available on the market since 2010.

Currently, four certification schemes for new buildings in the design phase have been developed according to building purposes: office buildings, residential buildings, family houses, and schools. The assessment framework for kindergartens is in progress, along with adjustments of current schemes for other project phases such as the operation (in use) phase and refurbishment (renovation).

Criteria for assessing the built environment are divided into groups: environmental, economic, and management criteria, plus the location of the building, including social criteria. Each criterion has an algorithm for setting up the indicator value, i.e., they are weighted according to importance. Criteria also differ according to the phase of the life cycle of a building (design, start of operation, operation).

2.4.2. Data Structure of the SBToolCZ Assessment Scheme for Residential Building

Detailed data analysis of the assessment tool SBToolCZ showed that more than 500 parameters are required as inputs for the complex quality and sustainability assessment of the residential building. The data sources differ according to the character of the information. The following data sources for the required information have been recognized:

- BIM—Building Information Model (202 parameters)—data linked to the objects and their semantic and classification, properties, or/and relationships.
- GIS (98 parameters)—data related to location, urban planning requirements, and building or urban zone protection, e.g., access to public transport, civic amenities, information of the locality, building location in specific or protected areas and zones (flooding zone, conservation areas, national parks, etc.).
- Documents and technical reports (128 parameters)—part of technical documentation but not part of the data model, such as energy consumption, energy sources, and internal microclimate quality (daylight, temperature, overheating, ventilation and air quality, acoustic parameters, etc.)
- Process description (48 parameters)—parameters describing the quality of processes in design, built-up, operational, and demolition phases, such as architectural competition, LCA and LCC analysis, and maintenance schedule.
- External sources (15 parameters)—parameters used for calculations and assessment, such as conversion factors, external databases of embodied emissions of used materials, and climatic data, such as precipitation amount, solar radiance, average external temperatures, and climatic load.

The BIM-based data model is a relevant source for almost half of the building parameters required for complex quality and sustainability assessment using SBToolCZ. The data structure of the model consists of the following data types:

- Direct or derived geometric properties—based on real of abstract/virtual 1D, 2D, 3D representations in the model:
    - Areas and surfaces—used as general project information, e.g., total floor space, total gross floor space, total heating volume, etc., or specifically for used methodology, e.g., green roofs area, green facades area, area for rainwater collection, external relaxing areas as balconies, terraces, etc.,
    - Lengths and dimensions—used as general project information, e.g., clear height or specifically for used methodology, e.g., dimensions of communication corridors for disabled users, etc.

    - Volumes—used as general project information as, e.g., the volume of used materials or specifically for used methodology, e.g., the volume of renewable and/or recycled materials, etc.
- Existence of a specific real or virtual item in the model predefined by the assessment methodology and number of those specific items used as general project information, e.g., number of users, number of dwelling units, parking places, or specifically for used methodology, e.g., number of parking places with charging station, number of bicycle parking places, etc.,

Properties of specific real or virtual objects of a distinct value predefined by specific enumeration types according to the assessment methodology, such as bicycle parking place being (i) non-guarded and non-covered, (ii) non-guarded and covered, (iii) covered and guarded, and (iv) covered and guarded by a central security system.

2.5. Connecting BIM and Complex Building Quality Assessment Method (CBQA)

2.5.1. Specifics of Complex Building Quality Assessment Methods

As shown in the previous chapter, the complexity of complex quality building assessment methods is compared to other calculations in the design phase much higher mainly because of the variety and diversity of required data. As mentioned in Section 2.4.2, in the case of SBToolCZ, only about 40% of required data can be exported from a BIM model.

About 60% of the information needed for the CBQA is still included in other sources, such as GIS, documents and technical reports, process description, or other external sources, and has to be mined manually.

A common BIM model corresponding to current BIM standards was used for the case study. The data structure of the model enables common data processing and analysis based on direct or derived geometric properties (areas and surfaces, lengths and dimensions, volumes).

The BIM model structure did not cover many parameters and data linked to the existence of a specific real or virtual item in the model predefined by the assessment methodology and number of those specific items, such as the number of parking places with charging stations, number of bicycle parking places, etc., and linked to properties of specific real or virtual objects of a distinct value predefined by specific enumeration types, e.g., non-guarded and non-covered, non-guarded and covered, covered and guarded, or covered and guarded bicycle parking place.

Some of those items were added to the model to use automatic data export, but still, some are missing, as the modeling was complicated and time-consuming compared to manual reading.

The fully automatic process for CBQA, under those circumstances, is currently not achievable, and the semi-automatic approach combining automatic data export with manual model reading shows a suitable workflow.

2.5.2. BIM Data Workflow for Complex Quality Building Assessment

The sustainability evaluation of a building or building project is a very complicated and time-consuming process and requires detailed information on building technologies, materials, energy systems, and processes and deep knowledge of the assessment methodology used for the assessment. Automation and the use of BIM in this process help to optimize the evaluation process and can improve the quality of the building project.

Considering current software possibilities and the complexity of quality assessment methods, the semi-automatic workflow described in the previous chapter was chosen as the appropriate strategy to combine BIM and complex quality assessment. Calculations and analysis for the complex building quality assessment are carried out outside the model in specific calculation tools, and the information needed for the assessment is exported in a predefined structure and can be automatically imported into the assessment software. A particular property or property set or entities that are not a common part of the data model are added manually in the calculation software.

2.5.3. BIM Data Modeling, and Use for Assessment Methodology

In the BIM-based data model, these parameters, such as lengths, dimensions, surfaces, areas, volumes, are gained from:

- 1D, 2D, and 3D building elements, such as walls, columns, slabs, windows, doors, and pipes.
- 1D, 2D, and 3D virtual elements—corridor width, room area, green roof area, green façade area, room volume, etc.

The assessment methodologies generally use:

- General data—required for every building project by national standards to prove the quality of the building, data and information used for the building permit process, and data used in the construction and operation phase of the building;
- Specific data—required for the assessment methodology or analysis required by the assessment methodology used for the building quality assessment.

The crucial part of the process enabling automation in the assessment process is the BIM model set up and its architecture according to the required information. The data structure of the assessment methodology is described above.

However, many different software applications enabling BIM are used in design practice; it is recommended to prepare the BIM model architecture based on IFC as an open and vendor-neutral data file format that allows cooperation, collaboration, and data use across different fields and specialization and using different software and tools for supporting analysis. The required data according to the data structure can be modeled by using the following IFC modeling tool:

- Direct or derived geometric properties
  - Areas—direct (or derived) attributes of IfcSpacialElements, such as spatial zone (IfcSpatialZone), space (IfcSpace), story (IfcBuildingStorey), building (IfcBuilding), and site (IfcSite).
  - Surfaces—direct (or derived) attributes of IfcElement as a generalization of all real or virtual components that make up an AEC product.
  - Lengths and dimensions—direct or derived attributes of IfcElement as a generalization of all real or virtual components that make up an AEC product or their annotations.
  - Volumes—direct (or derived) attributes of IfcProducts (IfcElement and/or IfcSpacialElements) or/and real or virtual components that make up an AEC product.
- Existence of a specific real or virtual item in the model predefined by the assessment methodology and number of those specific items represented by specific IfcProducts (IfcElement and/or IfcSpacialElements) (logical value 0/1 = true/false),
- Properties of specific real or virtual enumerated type of a distinct value modeled as appropriate IfcProducts (IfcElement and/or IfcSpacialElements) of a specific value.

3. Case Study—BIM for Complex Quality Building Assessment of a Residential Building

As a reference case study, the experimental residential building was used. The structural system (columns, shell, and horizontal slabs) consists of a high-performance reinforced concrete structure and is supplemented by timber substructures (façade, internal walls, and partitions). This system is mainly suitable for residential buildings, especially apartment houses, and it was developed in the project funded by the Technology Agency of the Czech Republic [47]. As a showcase, part of the apartment house was built in the experimental field of the University Centre for Energy Efficient Buildings (UCEEB) of the Czech Technical University in Prague (CTU) in 2019–2020. The model of the entire building and the real display are shown in Figure 5. The building itself consists of three independent apartments. Therefore, the whole project was fully under control within the whole design, construction, and operation process. The main goal of this project was described in conference papers [48]. A building information model of the TiCo project was used. The used model was prepared in Level of Development LOD350 [49].

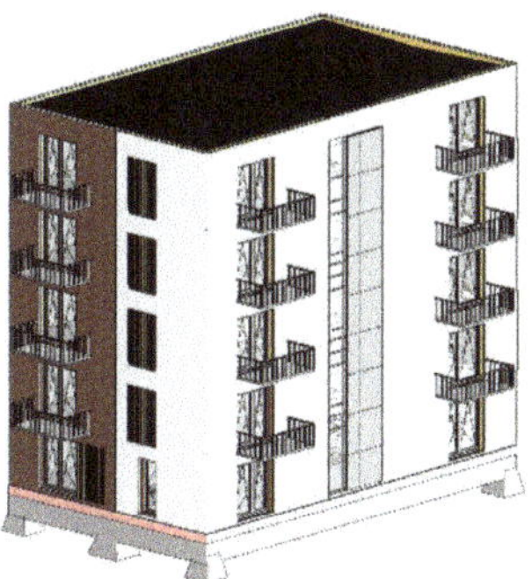

Figure 5. Model and real building of the TiCo project.

3.1. Geometric Data

For the purpose of this study, one information requirement from the social part of the SBToolCZ methodology is assessed: User comfort—the availability of a bike or stroller parking space. It is completed according to two tables (Tables 1 and 2). Each table assesses the stroller stands in the building from different perspectives. First, the table assesses the overall stroller stand solution:

Table 1. Number of credits according to the building design.

Item	Credits
No bike or stroller space	0
Uncovered space in exterior	2
Covered space in exterior	5
In the shared space inside the building	7
Individual garage stand inside the building (with enough space for bike or stroller)	10
Individual cellar (with enough space for bike or stroller)	10

Table 2. Minimum required bike or stroller spaces according to the number of the apartments.

Housing Unit Count	Minimum Required Bike or Stroller Spaces
<10	10
10 up to 29	20
30 up to 49	30
≥50	40

Additionally, there is a condition on the minimum storage floor area of bike or stroller based on the number of housing units.

The final credit count is calculated as the weighted average between all house units.

According to the IFC schema, we can complete an assessment of stroller stands by three different methods according to Sections 2.3.1–2.3.3. The utilized IFC data structure and semi-automatic workflow were selected. A description of part of the assessment is shown in Figure 6.

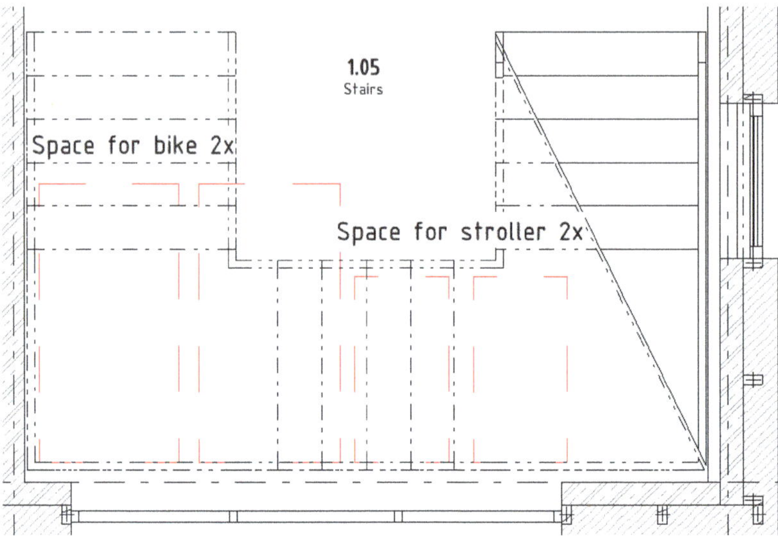

Figure 6. Bike and stroller spaces in the model.

Geometry is represented by a 2D or 3D object placed in the model. Their properties contain real values, which are compared with pre-set benchmarks, or they are used in the calculation.

3.2. Properties

3.2.1. Logical Value Type

The Boolean type logical value represents the values "true" or "false" and is used to automatically assess the presence of the specific predefined object in the model. If the object is placed in the model, it is possible to track the total numbers of predefined objects. In the sustainable quality assessment, this tool can be used to track the number of places for waste management, the number of parking places equipped with charging stations, etc.

This approach was used when an automated algorithm indicated that the model contains spaces for bikes and strollers.

3.2.2. Alphanumerical Value Type

This is the most commonly used type of parameter. It includes text values (description of all parameters), numbers (counts, areas, volumes, etc.), or any other property, such as material name, weight, and warranty.

In the proposed use case, the properties used for the description and number of elements are shown in Figure 7.

Availability of bike or stroller parking space	
Description	Count
Space for bike	2
Space for stroller	2

Figure 7. Schedule of spaces for bikes and strollers.

3.2.3. Enumeration Type

Specific properties of real or virtual objects can be defined according to a predefined structure following a quality assessment of specific criteria within the assessment framework. The enumeration type has distinct values that can be compared and assigned, e.g., roomy types, area types, façade types, etc. It is similar to a drop-down menu; this type of information is not possible to incorporate in the model (property of an element cannot contain more values from which the user can select one), and these values have to be written manually (or in a third-party software that allows using this functionality).

3.3. Data Processing

Complex quality assessment methodologies are more complicated than the standard LCA method already described in the literature and for which only the Bill of Quantities (BoQ) is necessary. A combination of all mentioned data types is needed for assessment; therefore, a more sophisticated data processing workflow is necessary. For SBToolCZ, an online application was developed. The whole assessment was completely online. The limitation is that so far, the application has only prepared the Excel tables. Therefore, it is not yet possible to import the IFC model directly. This is planned for version 2 of the application. The example with the TiCo project description is shown in Figure 8. The application can semi-automatically (with data prepared in the schedules) assess the building project, including different variants. It is suitable for the material and technical optimization of the project.

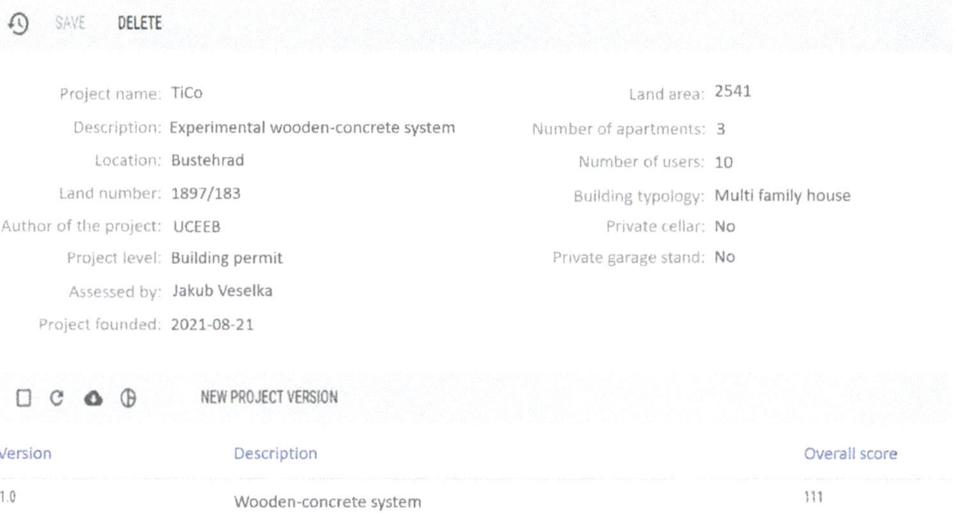

Figure 8. Description of the project in the online SBToolCZ application.

4. Discussion

4.1. The Data Structure of Assessment Schemes and Methods

This paper aims to analyze and verify possible BIM data-driven workflows for complex building quality assessment (CBQA) and a suitable BIM data structure set-up for automatic assessment and evaluation.

The development of most currently used assessment schemes and methodologies, such as LEED and BREEAM, including SBToolCZ, which is based on SBTool, started 15–20 years ago, and the structure of criteria and benchmarks was not created while considering the possibility of automatic assessment. Additionally, the IFC standards at that time were not established as a vendor-neutral and open system for data management. One of the problems that limit automated assessment is that the assessment methodologies do not follow the data structure used for automated processes.

Every single project and every single building are specific according to unique architectural design and structural, technological, and material solutions. One of the crucial aspects of assessment methodologies is to support the architectural quality, diversity, and originality of the architectural design of every building. This is the reason why many assessment criteria and benchmarks are often described in a continuous text, in examples of possible solutions that should motivate and which are not defined as finite enumeration.

For example, one of the quality parameters of a living unit is defined by the quality and size of external relaxation zones, such as balconies and loggias. It is impossible to make a finite enumeration of all spaces in the building that can be used for this purpose.

One of the missions for the future development of complex quality assessment methods is to adjust the criteria and benchmarks structure according to the data types, allowing a higher level of automatized data processing. In this process, it is fully recommended to consider the IFC data structure by adjusting semantic and classification of data objects, their properties, and relationships.

4.2. The Complexity of the Assessment Methodologies

The aim of using complex quality assessment methods is to cover wide aspects of building quality. It covers a wide range of criteria, including environmental quality and LCA, energy performance of buildings, durability, and resilience, quality of the internal environment, architectural and functional quality, and other social issues in considering the quality of locality. Due to this complexity, the assessment methods are very complicated and include many parameters from different branches. As mentioned above, SBToolCZ for residential buildings requires about 500 parameters for almost 35 criteria for the assessment. The data analysis of SBToolCZ proven within the case study shows that about 200 parameters can be derived from BIM and can be obtained based on current design standards, thus enriching the conclusions contained in the articles of Portuguese colleagues on the topic SBToolPT-H [18,36].

The question is if the ratio of BIM-based data can be higher than almost a half. According to the analysis of SBToolCZ data structure, when keeping the complexity of the assessment scheme, the ratio of data from other sources (GIS, technical reports, databases, processes) will still be very high. One option is to simplify the assessment methods, especially for use in the predesign of the early design phase, to use mainly BIM-based data for automatic assessment and optimization in this project phase. Another option is to adjust the structure of existing criteria in the assessment scheme so that it is better linked to the IFC data structure, and data mining from BIM will be easier.

4.3. BIM Data-Driven Workflows

Wastiels and Decuypere [36] divided BIM-LCA integration into five types: (i) integrating the tools by exporting the Bill of Quantities (BoQ) from the BIM environment into other tools; (ii) importing surfaces using the IFC format with predefined LCA profiles; (iii) processing information from a BIM tool in a BIM viewer tool, which is then transferred to dedicated LCA software; (iv) using specially developed plug-ins that enable LCA analysis within the BIM tool; (v) including the LCA information in the BIM objects that are used in the BIM model, instead of them being attributed to the building components in a later stage and in separate tools. As Obrecht et al. [13] state, the most adopted is the exchange of information via BoQ from BIM. As presented in the previous text and shown in the case study, the CBQA is more complex and requires a wide range of parameters, so using BoQ but also other mentioned BIM–LCA integration is not sufficient and does not cover all required data. According to this fact, different BIM data-driven workflows have to be used.

Using a semi-automatic workflow with calculations and analysis outside the data model, and data export directly from native model or via export to IFC or utilized database export, is a relevant approach and ensured higher level of automatic workflow, but still, the BIM model corresponds with current BIM standards.

Considering the complexity of the assessment methods and their fast development, on the one hand, and the complexity and diversity and development of standards and legislation in the building sector, on the other hand, the semi-automatic strategy brings many advantages. Using an independent calculation tool outside the data model enabling data import in predefined data structures from BIM allows better control over the assessment process and the BIM model and data management. These findings are also confirmed by colleagues from other leading universities [6,7,12].

To be able to process all the necessary steps, it is important to have the BIM in very good shape (e.g., classification system, proper model health, correct and precise naming, BoQ). These requirements must always be known at the beginning of the project and accepted by all stakeholders.

In a utilized IFC data structure and semi-automatic workflow, the amount of manual assessment is significantly lower than in low structured model data and manual workflow. Thus, to be able to achieve a high degree of automation, it is necessary to move the whole BIM environment to a highly structured model data and fully automatic workflow. There are many obstacles and challenges (e.g., lack of legislation, low availability of skilled manpower), but the direction for the BIM environment to the higher level of automation and usage of robotics seems to be clear [50].

A literature review showed that specifically for the Czech Republic, SBToolCZ methodology is not digitalized, and legislation and standardization of the BIM are not finished; therefore, it is not yet possible to use highly structured model data and fully automatic workflow. These conclusions aim to advance, not only in the Czech Republic, the discussion regarding the necessary completion of the above-mentioned standardization process.

4.4. Level of Information of BIM

As was mentioned before, the SBToolCZ methodology needs around 500 inputs for the full assessment, and only around 200 of them can be taken relatively easily from the BIM. Due to the fact that the IFC format is very capable in terms of storing data, all the other parameters can be stored directly in the model as well. However, it is not clear how it would help to speed up the process of assessment. Currently, the IFC format is being developed rapidly, and no software can use its full potential of it. The only option to use all the possibilities is manual data import in the IFC file, which is inefficient. Further research showing the optimal amount of data taken out of the model is necessary for every mentioned complex quality assessment methodology. Similar conclusions regarding the IFC format have been reached in other scientific studies [41,43,45]. It is a question of how long this state will take before there is automation between individual data formats.

5. Conclusions

Using Building Information Modeling for complex building design and optimizing building quality in a design phase becomes an important part of integrated and advanced building design. Using data from an information model in the design phase allows efficient assessment of different design strategies and structural variants and a higher quality of final design. This paper aimed to analyze and verify possible BIM data-driven workflows for Complex Building Quality Assessment (CBQA) and a suitable BIM data structure set-up for an automatic assessment and evaluation. For an efficient automation process in complex quality assessment of the building in the design phase, it is necessary first to understand the data structure of the Industry Foundation Classes, which is widely accepted and used for buildings, and second, to understand the data structure of the assessment methodology used for the specific assessment. This article describes the potential of the automatic assessment of complex buildings based on the experience gained in a project where the complex building quality was tested on a real pilot project of a residential building using SBToolCZ, the Czech national assessment method. The outcomes present the experience with semi-automatic data workflow and recommendations for setting up the data model of a building for automatic assessment. The outcomes also present the experience gained by combining BIM and automatic complex building quality assessment based on semi-automatic data workflow combined with data export from a structured BIM model with the independent calculation tool. The case study proved this method as an efficient approach to automatic assessment.

Author Contributions: Conceptualization, J.R. and J.V.; methodology, Z.R.; investigation, S.V.; resources, S.V.; writing—original draft preparation, J.R., J.V., Z.R. and S.V.; writing—review and editing, J.V.; supervision, P.H.; project administration, J.V.; funding acquisition, J.R. All authors have read and agreed to the published version of the manuscript.

Funding: This research was funded by the Technology Agency of the Czech Republic, grant number TN01000056, research project NCK CAMEB Centre for Advanced Materials and Efficient Buildings.

Institutional Review Board Statement: Not applicable.

Informed Consent Statement: Not applicable.

Data Availability Statement: Not applicable.

Acknowledgments: The research outcomes have been achieved within the research project BIMIP Building Information Modelling—Integrated Whole Life Cycle Process in the framework of the research project number TN01000056 NCK CAMEB Centre for Advanced Materials and Efficient Buildings.

Conflicts of Interest: The authors declare no conflict of interest.

References

1. Eastman, C.; Teicholz, P.; Sachs, R.; Liston, K. *BIM Handbook: A Guide to Building Information Modeling for Owners, Managers, Designers, Engineers and Contractors*; Wiley: New York, NY, USA, 2011; ISBN 9780470541371.
2. Abraham, F.; Cahn, J.; Fitzerald, B.; Greenberg, E. *Strategy in the Face of Disruption: A Way Forward for the North American Building-Products Industry*; McKinsey & Company; Available online: https://www.mckinsey.com/industries/private-equity-and-principal-investors/our-insights/strategy-in-the-face-of-disruption-a-way-forward-for-the-north-american-building-products-industry (accessed on 30 December 2021).
3. Barbosa, F.; Woetzel, J.; Mischke, J.; Ribeirinho, M.J.; Sridhar, M.; Parsons, M.; Bertram, N.; Brown, S. *Reinventing Construction: A Route to Higher Productivity*; Mckinsey Global Insititute; 2017; p. 168. Available online: http://www.mckinsey.com/industries/capital-projects-and-infrastructure/our-insights/reinventing-construction-through-a-productivity-revolution (accessed on 30 December 2021).
4. Prebanić, K.R.; Vukomanović, M. Realizing the need for digital transformation of stakeholder management: A systematic review in the construction industry. *Sustainability* **2021**, *13*, 12690. [CrossRef]
5. Pérez-García, A.; Martín-Dorta, N.; Aranda, J.Á. BIM Requirements in the Spanish Public Tender—Analysis of Adoption in Construction Contracts. *Buildings* **2021**, *11*, 594. [CrossRef]
6. Fang, T.; Zhao, Y.; Gong, J.; Wang, F.; Yang, J. Investigation on maintenance technology of large-scale public venues based on bim technology. *Sustainability* **2021**, *13*, 7937. [CrossRef]
7. Acampa, G.; Diana, L.; Marino, G.; Marmo, R. Assessing the transformability of public housing through bim. *Sustainability* **2021**, *13*, 5431. [CrossRef]
8. Ren, G.; Li, H.; Zhang, J. A BIM-based value for money assessment in public-private partnership: An overall review. *Appl. Sci.* **2020**, *10*, 6483. [CrossRef]
9. Andújar-Montoya, M.D.; Galiano-Garrigós, A.; Echarri-Iribarren, V.; Rizo-Maestre, C. BIM-LEAN as a methodology to save execution costs in building construction-An experience under the spanish framework. *Appl. Sci.* **2020**, *10*, 1913. [CrossRef]
10. Altohami, A.B.A.; Haron, N.A.; Ales@Alias, A.H.; Law, T.H. Investigating approaches of integrating BIM, IoT, and facility management for renovating existing buildings: A review. *Sustainability* **2021**, *13*, 3930. [CrossRef]
11. Hollberg, A.; Lützkendorf, T.; Habert, G. Top-down or bottom-up?—How environmental benchmarks can support the design process. *Build. Environ.* **2019**, *153*, 148–157. [CrossRef]
12. Hollberg, A.; Genova, G.; Habert, G. Evaluation of BIM-based LCA results for building design. *Autom. Constr.* **2020**, *109*, 102972. [CrossRef]
13. Obrecht, T.P.; Röck, M.; Hoxha, E.; Passer, A. BIM and LCA integration: A systematic literature review. *Sustainability* **2020**, *12*, 5534. [CrossRef]
14. Manzoor, B.; Othman, I.; Gardezi, S.S.S.; Harirchian, E. Strategies for adopting building information modeling (Bim) in sustainable building projects—A case of Malaysia. *Buildings* **2021**, *11*, 249. [CrossRef]

15. Zhang, L.; Chu, Z.; He, Q.; Zhai, P. Investigating the constraints to buidling information modeling (BIM) applications for sustainable building projects: A case of China. *Sustainability* **2019**, *11*, 1896. [CrossRef]
16. Akhanova, G.; Nadeem, A.; Kim, J.R.; Azhar, S.; Khalfan, M. Building information modeling based building sustainability assessment framework for Kazakhstan. *Buildings* **2021**, *11*, 384. [CrossRef]
17. Manzoor, B.; Othman, I.; Kang, J.M.; Geem, Z.W. Influence of building information modeling (Bim) implementation in high-rise buildings towards sustainability. *Appl. Sci.* **2021**, *11*, 7626. [CrossRef]
18. Carvalho, J.P.; Almeida, M.; Bragança, L.; Mateus, R. Bim-based energy analysis and sustainability assessment—Application to portuguese buildings. *Buildings* **2021**, *11*, 246. [CrossRef]
19. Barone, G.; Buonomano, A.; Forzano, C.; Giuzio, G.F.; Palombo, A. Improving the efficiency of maritime infrastructures through a bim-based building energy modelling approach: A case study in Naples, Italy. *Energies* **2021**, *14*, 4854. [CrossRef]
20. Deng, M.; Tan, Y.; Singh, J.; Joneja, A.; Cheng, J.C.P. Computers in Industry A BIM-based framework for automated generation of fabrication ¸ade panels drawings for fac. *Comput. Ind.* **2021**, *126*, 103395. [CrossRef]
21. Zima, K.; Mitera-Kiełbasa, E. Employer's information requirements: A case study implementation of bim on the example of selected construction projects in Poland. *Appl. Sci.* **2021**, *11*, 10587. [CrossRef]
22. Jang, K.; Kim, J.W.; Ju, K.B.; An, Y.K. Infrastructure BIM platform for lifecycle management. *Appl. Sci.* **2021**, *11*, 10310. [CrossRef]
23. Lesniak, A.; Monika, G.; Skrzypczak, I. Barriers to BIM Implementation in Architecture, Construction and Engineering Projects—The Polish Study. *Energies* **2021**, *14*, 2090. [CrossRef]
24. Hammad, A.W.A.; da Costa, B.B.F.; Soares, C.A.P.; Haddad, A.N. The Use of Unmanned Aerial Vehicles for Dynamic Site Layout Planning in Large-Scale Construction Projects. *Buildings* **2021**, *11*, 602. [CrossRef]
25. Zanni, M.; Sharpe, T.; Lammers, P.; Arnold, L.; Pickard, J. Developing a methodology for integration of whole life costs into BIM processes to assist design decision making. *Buildings* **2019**, *9*, 114. [CrossRef]
26. Edirisinghe, R.; Pablo, Z.; Anumba, C.; Tereno, S. An Actor–Network Approach to Developing a Life Cycle BIM Maturity Model (LCBMM). *Sustainability* **2021**, *13*, 13273. [CrossRef]
27. Tan, Y.; Li, S.; Liu, H.; Chen, P.; Zhou, Z. Automation in Construction Automatic inspection data collection of building surface based on BIM and UAV. *Autom. Constr.* **2021**, *131*, 103881. [CrossRef]
28. Chen, K.-L.; Tsai, M.-H. Conversation-Based Information Delivery Method for Facility Management. *Sensors* **2021**, *21*, 4771. [CrossRef]
29. Mannino, A.; Dejaco, M.C.; Re Cecconi, F. Building information modelling and internet of things integration for facility management-literature review and future needs. *Appl. Sci.* **2021**, *11*, 3062. [CrossRef]
30. Pavón, R.M.; Alvarez, A.A.A.; Alberti, M.G. Possibilities of bim-fm for the management of Covid in public buildings. *Sustainability* **2020**, *12*, 9974. [CrossRef]
31. *ISO 14044:2006*; Environmental Management—Life Cycle Assessment—Requirements and Guidelines. Available online: https://www.iso.org/standard/38498.html (accessed on 30 December 2021).
32. Frischknecht, R.; Birgisdottir, H.; Chae, C.-U.; Lützkendorf, T.; Passer, A.; Alsema, E.; Balouktsi, M.; Berg, B.; Dowdell, D.; García Martínez, A.; et al. Comparison of the environmental assessment of an identical office building with national methods. *IOP Conf. Ser. Earth Environ. Sci.* **2019**, *323*, 012037. [CrossRef]
33. Frischknecht, R.; Ramseier, L.; Yang, W.; Birgisdottir, H.; Chae, C.U.; Lützkendorf, T.; Passer, A.; Balouktsi, M.; Berg, B.; Bragança, L.; et al. Comparison of the greenhouse gas emissions of a high-rise residential building assessed with different national LCA approaches—IEA EBC Annex 72. *IOP Conf. Ser. Earth Environ. Sci.* **2020**, *588*, 022029. [CrossRef]
34. Cavalliere, C.; Hollberg, A.; Dell'Osso, G.R.; Habert, G. Consistent BIM-led LCA during the entire building design process. *IOP Conf. Ser. Earth Environ. Sci.* **2019**, *323*, 012099. [CrossRef]
35. Hollberg, A.; Kiss, B.; Röck, M.; Soust-Verdaguer, B.; Wiberg, A.H.; Lasvaux, S.; Galimshina, A.; Habert, G. Review of visualising LCA results in the design process of buildings. *Build. Environ.* **2021**, *190*, 107530. [CrossRef]
36. Carvalho, J.P.; Alecrim, I.; Bragança, L.; Mateus, R. Integrating BIM-Based LCA and Building Sustainability Assessment. *Sustainability* **2020**, *12*, 7468. [CrossRef]
37. Wastiels, L.; Decuypere, R. Identification and comparison of LCA-BIM integration strategies. *IOP Conf. Ser. Earth Environ. Sci.* **2019**, *323*, 012101. [CrossRef]
38. Veselka, J.; Nehasilová, M.; Dvořáková, K.; Ryklová, P.; Volf, M.; Růžička, J.; Lupíšek, A. Recommendations for Developing a BIM for the Purpose of LCA in Green Building Certifications. *Sustainability* **2020**, *12*, 6151. [CrossRef]
39. Vázquez-Rowe, I.; Córdova-Arias, C.; Brioso, X.; Santa-Cruz, S. A method to include life cycle assessment results in choosing by advantage (Cba) multicriteria decision analysis. A case study for seismic retrofit in peruvian primary schools. *Sustainability* **2021**, *13*, 8139. [CrossRef]
40. Vonka, M.; Lupíšek, A.; Hájek, P. SBToolCZ: Sustainability Rating System in the Czech Republic. Available online: https://www.researchgate.net/publication/263209319_SBToolCZ_Sustainability_rating_system_in_the_Czech_Republic (accessed on 30 December 2021).
41. *ISO 16739-1:2018*. Industry Foundation Classes (IFC) for Data Sharing in the Construction and Facility Management Industries—Part 1: Data Schema. Available online: https://www.iso.org/standard/70303.html (accessed on 30 December 2021).
42. Gerbino, S.; Cieri, L.; Rainieri, C.; Fabbrocino, G. On BIM Interoperability via the IFC Standard: An Assessment from the Structural Engineering and Design Viewpoint. *Appl. Sci.* **2021**, *11*, 11430. [CrossRef]

43. Gkeli, M.; Potsiou, C.; Soile, S.; Vathiotis, G.; Cravariti, M.-E. A BIM-IFC Technical Solution for 3D Crowdsourced Cadastral Surveys Based on LADM. *Earth* **2021**, *2*, 605–621. [CrossRef]
44. Horn, R.; Ebertshäuser, S.; Di Bari, R.; Jorgji, O.; Traunspurger, R.; von Both, P. The BIM2LCA approach: An industry foundation classes (IFC)-based interface to integrate life cycle assessment in integral planning. *Sustainability* **2020**, *12*, 6558. [CrossRef]
45. Tang, S.; Shelden, D.R.; Eastman, C.M.; Pishdad-bozorgi, P.; Gao, X. Automation in Construction BIM assisted Building Automation System information exchange using BACnet and IFC. *Autom. Constr.* **2020**, *110*, 103049. [CrossRef]
46. Industry Foundation Classes (IFC). 2021. Available online: https://technical.buildingsmart.org/standards/ifc/ (accessed on 7 March 2021).
47. Ryklová, P.; Mančík, Š.; Lupíšek, A. Environmental Benefits of Timber-Concrete Prefabricated Construction System for Apartment Buildings—A Simplified Comparative LCA Study. *IOP Conf. Ser. Earth Environ. Sci.* **2019**, *290*, 012083. [CrossRef]
48. Veselka, J.; Růžička, J.; Lupíšek, A.; Hájek, P.; Mančík, Š.; Žd'ára, V.; Široký, M. Connecting BIM and LCA: The Case Study of an Experimental Residential Building. *IOP Conf. Ser. Earth Environ. Sci.* **2019**, *323*, 012106. [CrossRef]
49. BIMFORUM—Level of Development (LOD). Available online: https://bimforum.org/lod/ (accessed on 30 December 2021).
50. Zhang, J.; Luo, H.; Xu, J. Towards fully BIM-enabled building automation and robotics A perspective of lifecycle information flow. *Comput. Ind.* **2022**, *135*, 103570. [CrossRef]

MDPI

Article

Reverse Logistics Performance Indicators for the Construction Sector: A Building Project Case

Mochamad Agung Wibowo [1], Naniek Utami Handayani [2,*], Anita Mustikasari [3], Sherly Ayu Wardani [2] and Benny Tjahjono [4]

1 Department of Civil Engineering, Diponegoro University, Semarang 50275, Indonesia; agung.wibowo@ft.undip.ac.id
2 Department of Industrial Engineering, Diponegoro University, Semarang 50275, Indonesia; sherlyayuwardani@alumni.undip.ac.id
3 Department of Management, Yogyakarta State University, Yogyakarta 55281, Indonesia; anita.mustikasari@uny.ac.id
4 Centre for Business in Society, Coventry University, Coventry CV1 5FB, UK; benny.tjahjono@coventry.ac.uk
* Correspondence: naniekh@ft.undip.ac.id

Abstract: While the performance evaluation of reverse logistics (RL) practices in the construction sector is crucial, it is seemingly limited compared to that in the manufacturing sector. As the project life cycle in the construction sector is typically long, effective coordination among the stakeholders is needed to integrate RL into each phase of the project life cycle. This paper proposes a new model of RL for the construction industry, incorporating the dimensions, elements, and, most importantly, indicators needed for the evaluation of RL performance. The model was initially derived from the extant literature. It was then refined through (1) focus group discussion, by which suggestions pertinent to the proposed model were collated from academics and practitioners, and (2) judgments by academics and practitioners to validate the model. The validated model includes 21 indicators to measure RL performance, spanned throughout the green initiation, green design, green material management, green construction, and green operation and maintenance phases. The paper offers a new method for how RL can be adopted in the construction industry by proposing an innovative model that will benefit stakeholders in the construction industry.

Keywords: performance evaluation; project life cycle; reverse logistics; construction; indicators

Citation: Wibowo, M.A.; Handayani, N.U.; Mustikasari, A.; Wardani, S.A.; Tjahjono, B. Reverse Logistics Performance Indicators for the Construction Sector: A Building Project Case. *Sustainability* **2022**, *14*, 963. https://doi.org/10.3390/su14020963

Academic Editor: Antonio Garcia-Martinez

Received: 2 December 2021
Accepted: 8 January 2022
Published: 15 January 2022

1. Introduction

In recent years, environmental problems have become a serious issue in construction projects. The construction industry generates a significant amount of waste, which may have negative impacts on society and the environment [1,2]. Construction waste can typically be categorized as solid waste (e.g., garbage, mud, air pollution, and CO_2 emissions) and non-solid waste (e.g., delay, rework, and over costing during the construction process) [2]. According to a report by the United Nations Environment Program (UNEP), the building sector accounts for up to 40% of global annual energy consumption and 20% of global annual water usage and contributes 40% of global annual total waste as a result of building construction and demolition activities [3]. Urbanization has increased the demand for buildings and infrastructure, which in turn leads to the consumption of material resources, water, and energy and generates large quantities of material waste throughout the project's lifespan [4]. For example, Surahman et al. [4] reported that the volumes of demolition debris and waste in a major city in Indonesia (Jakarta) reached approximately 123.9 million tons between 2012 and 2020, all of which went to landfills. The production of Hebel light bricks used as constituent materials for building projects also generates 6.88% (4021.8 m^3) of waste per month [4].

Against this backdrop, the strategies proposed in the extant literature have been mostly geared toward improving construction supply chain management (SCM) by minimizing waste and adding value by conducting effective stewardship of information and refining logistics [5–7]. Green SCM (GSCM) aims to manage construction business processes in a more environmentally friendly manner [8,9]. GSCM in the construction sector typically follows a project life cycle (PLC) that includes green initiation, green design, green materials management, green construction, and green operations and maintenance (O/M) phases [10].

Reverse logistics (RL) is a subset of GSCM. Reusing, recycling, and remanufacturing are considered to be RL functions that ensure the attainment of GSCM [8]. Implementing RL is regarded as a "remedial" measure that moderates the detrimental impacts of construction projects on the environment and enables construction industries to be more efficient, gaining economic benefits and sustainable competitiveness [11,12]. RL aims to recover waste generated by construction activities, simultaneously maximizing the retained value of construction materials and reducing the costs of waste management [12,13].

However, RL appears to be implemented less frequently in the construction sector than in the manufacturing sector. One reason for this deficiency is the fact that the product life cycle in construction is generally long—much longer than in the manufacturing sector. Unlike in the manufacturing industry, where RL is typically well integrated and considered from the beginning of the product development stage, RL in the construction industry has been treated as an independent activity. Coordination between stakeholders is therefore critical to integrate RL into each phase of the PLC [14]. In this way, the design practice for deconstruction would allow a systematic demolition of buildings conducted in such a way that the demolition materials remain high in value and the amount of material damage is reduced. To maximize RL in the construction sector, construction practitioners require the awareness and know-how to incorporate RL concepts (values) from the initiation phase [12]. This step must be supported by an adequate capacity in the construction sector to evaluate the performance of RL practices [15]. Hosseini et al. [16] conducted one such study of RL practices in the construction sector, while Farida et al. [17] incorporated RL to measure the performance of green construction. Pushpamali et al. [12] attempted to incorporate RL into various decisions made by the project owners at the preconstruction stage; however, arguably their work only provided a conceptual scheme of RL decisions in construction. Finally, Hammes et al. [15] developed a measurement tool for RL performance during the construction phase carried out by the contractor, involving supplies, internal logistics, and waste management.

The study discussed in this paper focuses on the development of a performance measurement system for RL in accordance with the constructions' PLCs. The paper also proposes a new model of RL in the construction industry, along with the dimensions, elements, and indicators for the evaluation of RL performance throughout the PLC. The contributions may offer substantial benefits for stakeholders in the construction industry related to coordination and collaboration.

The remainder of the paper is structured as follows. We first review the literature on waste in the construction industry, green SCM, RL in manufacturing and construction, and performance evaluations of RL. Based on the literature review, we conceptualize the performance evaluation of RL in construction. This conceptualization provides a research framework related to the theme design and conceptual relationships. We then proceed to develop and examine the measurement of RL performance in the construction industry through focus group discussion (FGD) and expert judgments. We conclude with the results and discuss the theoretical contributions and practical implications of the research.

2. Literature Review

2.1. RL Concept and Applications

RL is traditionally triggered by the need for product returns in retail sectors [18]; manufacturers may return raw materials to their suppliers because they are of poor quality, in excess or in surplus in another way, unused, or out-of-specification. Manufacturers may also recall their products, such as car braking systems, due to manufacturing defects, commercial returns, unsold out-of-season products, or wrong deliveries. Finally, in many countries, customers have the right to return items because they are unwanted or, according to warranty, at the end of life or end of service.

Economically, companies that choose to carry out RL activities are motivated by the opportunity to recover resources cheaply and add value by transforming them into other resources with higher commercial values. Due to growing competition, many companies are forced to take back and offer refunds for unwanted products from their customers. Other companies act in strategically risk-averse ways by preventing their products or critical components from leaking to their competitors or secondary markets. With the wide spread of product-service system (PSS) business models, many companies sell products as part of their service offering (leasing) and consequently have to take the products (or assets) off the field for service/maintenance and repair [19]. Finally, the regulations and laws pertinent to environmental consciousness, such as extended producer responsibility and the "right to repair law", place extra pressure on manufacturers to adhere strictly to public environmental policy.

While the scope and definitions of RL were initially somewhat limited to the movement of products in the opposite direction to forward logistics [20–22], focus has now shifted to activities within the reverse flow, such as component recovery, reuse, and recycling. RL is gaining the attention of industrialists and academic researchers due to the enormous quantity of waste generation in manufacturing and construction sectors, which is leading to increased environmental pressure [23]. In an expansion of its initial definition, RL now incorporates the process of planning, implementing, and controlling efficiently and effectively the reuse of disposed products [24]. This wider notion largely echoes the classic proposition of Rogers and Tibben-Lembke [25], who extended the definition of RL given by the Council of Logistics Management (now Council of Supply Chain Management Professionals (CSCMP)) to emphasize "the flow of raw materials, in-process inventory, finished goods, and related information from the point of consumption to the point of origin for the purpose of recapturing value or proper disposal" [25]. Depending on the various underlying motivations, the detailed structure of RL can vary to include activities such as distribution, sorting, reselling, refurbishment, remanufacturing, recycling, and disposal, among others, with the ultimate aim of recapturing the value of products after the point of sales and/or after the end of useful life [9].

Many studies consider RL from the moment the waste is generated and must be sent for recycling or environmentally correct disposal [26]. However, Guarnieri et al. [27] emphasize that RL must be considered for the entire product life cycle, including the planning and design of the productive process. The management of the product life cycle needs industrial synergies within large-scale networks to collect, recycle, reuse, and recover end-of-life products [28]. RL in the manufacturing industry would close the loop of the supply chain at different points, resulting in reusing the products as entire products, modules, or a combination of modules and materials [29].

2.2. RL in the Construction Industry

In this study, RL in the construction industry is defined as the process of planning, practicing, and managing construction items and material flows [16]. It involves information flow for effective construction waste and disposal management in the PLC [10]. The configuration and quantity of building sectors' waste are related to the waste's recycling potential, which is critical to closing material loops and reducing waste and emissions in a circular economy [30].

There is a fundamental difference between the RL concept in the manufacturing and construction sectors. This is due to the difference in the main source of returned items and the stage at which they become available. In the construction industry, major parts of materials become available after the end of life of a building, which may take a long time. This time factor may impede the implementation of RL in many ways in the construction sector and highlights the need to conceptualize RL for particular use within the construction industry due to the observed discrepancies in the associated processes between the manufacturing and construction contexts [16].

In general, RL in construction can be categorized according to the following dimensions: demolition, component recovery, reuse [12,16], deconstruction [31], and recycling [12,16]. Demolition waste is defined as a mixture of surplus materials generated from construction, renovation, and demolition activities [32]. Component recovery involves the reuse of secondary resources instead of recycling [33]. Reuse is the activity of reusing materials without the need for additional processes. Design for deconstruction (DFD) is an approach related to reusing building materials or components that have high durability [31]. Recycling is the activity of reprocessing a material to obtain material of the same quality [16].

Previous studies discussing RL in construction have been limited to individual, specific phases [12,14] due to a lack of knowledge regarding RL and initiating designs that make deconstruction impossible [14]. The deconstruction process becomes difficult to carry out at the end of life of the project if, from its beginning, the project has not been designed using the DFD concept [14]. DFD is an essential strategy when producing a modular product that aims to develop a building with a design that has high durability and easy-to-use materials in the end-of-life phase [31].

The integration of end-of-life strategies into the initiation phase is also critical for successful RL implementation in construction because the amount of material that can be recovered at the end of the building's life is determined by the type and quality of materials used in the new construction. Therefore, RL concepts should ideally be taken into consideration at an early decision (initiation) phase to allow for the collection of recovered materials to be properly managed [12]. An environmentally friendly building that is efficient throughout its life cycle (conception, design, construction, maintenance, and demolition) offers ways to reducing environmental impacts. It can provide more efficient and effective use of materials, water, and energy, thus maximizing the retained value of the construction materials while reducing the costs of waste management [12,15,34,35].

Previous research suggests that RL frameworks developed for the manufacturing industry would equally be effective in other contexts, including the construction industry [36]. For instance, the scenario analysis conducted by Surahman et al. [4] for RL material flows in the building sector would decrease final waste disposal by more than 90%. RL has also been reported to have reduced costs related to the transportation of construction materials by 25% [37]. It has been argued, therefore, that launching RL within a project environment can add value to a construction business [1]. RL, according to the construction literature, could eliminate risks and uncertainties [38], resulting in cost reduction [39] and boosting the efficiency level of the RL system through cooperation between stakeholders involved in the construction industry [12]. This would also reduce the costs of inventory, transportation, and waiting time, indirectly facilitating the minimization of waste within the system. It would also potentially improve the industry's awareness of the benefits of RL, which may result in an increased level of support from top management [40].

This research focuses on the implementation of RL in so-called "closed-loop constructions", in which the processed materials are immediately reused so that the amount of waste is minimized. In past decades, construction and demolition waste (C&DW) was mostly used for road foundations and embankments, which was considered downcycling [41]. However, in recent years, recycling C&DW as aggregates in new concrete has drawn significant attention, with similar interest shown in recycled waste glass or asphalt shingle as a raw material in the manufacture of cement [42]. Previous research has suggested that construction practitioners should give further attention to improving the management of concrete, masonry (bricks and concrete/stone blocks), mortar, and ceramic wastes because these four types of C&DW have the largest potential for recycling [43,44]. In a case study in China, Yuan et al. [45] claimed that the major obstacles in C&DW management were the lack of a well-developed waste recycling market, insufficient regulatory support, and the trend in building designs paying insufficient attention to waste reduction. A similar situation can be found in Indonesia, where stakeholders in the pre-construction phase, such as building owners and design consultants, lack knowledge about how to apply RL in the building construction process [10]. Hence, in the initiation phase, the building owner plays a vital role in creating/building environmentally friendly value by applying RL to the planned construction. Furthermore, the RL concept should be realized in the detailed engineering design (DED) made by the design consultants to facilitate DFD. When these early phases are skipped, the deconstruction process becomes hard to achieve, making the RL implementation in building projects unproductive [14].

2.3. *Performance Measures of RL in the Construction Industry*

According to Badenhorst [46], it is essential that companies manage all the processes involved in RL efficiently and effectively so that they understand all its aspects. The purpose of performance evaluation in RL is to measure the efficiency and effectiveness of the activities involved in the materials' reverse flow to assess whether these activities can be improved and where it is necessary to invest more resources to increase their benefit [47].

An example of such RL performance evaluation is the ten key performance indicators (KPIs) endorsed by the New Zealand (NZ) government, which address both project and company performances in the construction sector [48]. Although the NZ government intends to endorse a broad set of practical indicators, there is no appropriate KPI to measure logistics performance in the construction industry that is especially pertinent to RL. For performance measurement to be effective, there are several criteria for selecting a KPI. First, a KPI can translate practices and measures into practical knowledge and make it possible to identify and adopt superior performance standards [49]. Performance measures are also used to measure and improve the efficiency and quality of the process and identify opportunities for progressive improvements in process performance. A KPI should be able to measure and monitor the practice, as well as address the characteristics of construction projects that involve many tiers of practitioners on site [50].

Several studies have examined the measurement of RL performance in constructions [15]. Hammes et al. [15] stated several aspects to compare in building RL performance models in the construction industry. However, they focused only on RL performance assessment in the construction phase, which concerns supplies (green purchasing), internal logistics (use of materials, reuse of material, return of investment, and customer satisfaction), and waste management (storage, transportation, and awareness of workers in waste management). Furthermore, the study did not consider the involvement of stakeholders in measuring RL performance; to achieve success in a project, it is important to unify the understanding and perceptions of stakeholders when carrying out the project. For example, Pushpamali et al. [12] found that the role of stakeholders is vital in RL implementation. Pushpamali et al. [12] also stated that in the construction industry, the impact of upstream activities is more substantial than that of their end-of-life counterparts, and the initiation phase is particularly important for successful RL implementation. Therefore, the model used to measure RL performance, which should be integrated into the PLC throughout the initiation, design, material management, construction, commissioning and handover, and O/M phases, needs to be more efficient than that when the measurement is done separately [11,12].

The RL development model in this study was adapted from the scheme of Pushpamali et al. [12] and the GSCM concept in the construction sector developed by Wibowo et al. [10]. This model was also evolved in relation to the concept, dimensions, elements, and indicators of each phase of the PLC through an interview process and FGD with respondents (academic researchers and practitioners) as well as through the literature review. The development framework used to measure RL performance can be seen in Figure 1.

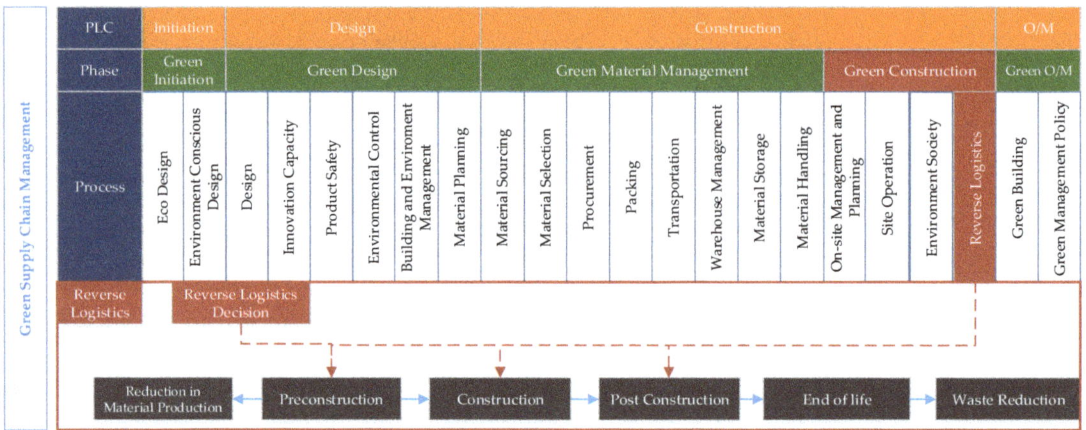

Figure 1. Framework for measuring RL performance (adapted from Wibowo et al. and Pushpamali et al.).

In the construction sector, the implementation of RL starts from the upstream supply chain, which represents all activities before the development process or preconstruction [12]. This development framework also integrates with the GSCM-PLC system, including the green initiation, green design, and green material management phases. The downstream represents all activities carried out after the construction process (postconstruction), such as the green O/M phase and end of life, including waste management and demolition activities [12].

RL should be integrated into the PLC system. However, RL is currently carried out only during the construction phase, or material is recycled after the construction phase. For instance, recycled material, such as the remainder of a cast, will be reused as material for lighter structural work, such as curbs or parking stoppers, in the construction phase. Based on these observations, improvements are needed. Such improvement needs to begin with the measurement of RL performance based on the PLC from the initiation, design, construction, commissioning and handover, and O/M phases to determine the improvement starting point precisely.

Performance measurement tools related to RL have been created in the manufacturing industry. Shaik and Abdul-Kader [51] developed a measurement tool called the overall comprehensive performance index (OCPI), which relates to aspects of financial, process, stakeholder, and innovative perspectives in manufacturing. Bansia et al. [52] also measured RL performance according to financial, customer, internal business, and innovation and growth aspects. Guimarães and Salomon [53] examined the level of urgency of indicators in the implementation of RL, considering recapture value, operation cost, technological innovation, encouragement of recycling, social and environmental acts, employment creation, long-term relationship, differentiated service, and compliance with legislation. Morgan et al. [54] looked at the effect of stakeholder commitment to implementing RL on the company's operational performance through variables, commitment to implementing a sustainable supply chain, commitment to implementing RL, sustainable RL capability, and operational performance.

3. Research Method

The purpose of the study described in this paper is to offer a new perspective on how RL can be adopted in the construction industry from the initiation, design, materials management, and construction phases to the O/M phases. The paper also proposes a new model of RL for the construction industry, along with the dimensions, elements, and, more importantly, indicators for the evaluation of RL performance during the construction's PLC.

The method adopted in this study consists of three major steps: (1) desk-based research to propose the initial RL measurement indicators, (2) FGD to collate suggestions from academics and practitioners regarding the indicators proposed and (3) validation of the indicators, also by academics and practitioners.

First, the proposed performance indicators gathered from the literature were distributed to academics and practitioners from the construction industry via an open questionnaire, which allowed respondents to make recommendations or suggestions about indicators that should be added. It was hoped that this would not only improve the accuracy but also ensure the practicality and completeness of the indicators. The respondents consisted of three academics and 13 practitioners from the construction sector. The 13 practitioners involved in project appraisal were split on the basis of their roles in each phase of the research considering the criteria proposed by Etikan et al. [55], but the academics, whose research focused on green design, RL, and sustainable constructions, partook in assessments of all the phases. These phases comprised the following:

1. Green initiation phase. The respondents who assessed RL performance at this phase were typically project owners as they were able to assess commitment to implementing RL in a construction project.
2. Green design phase. In this phase, the performance assessment was carried out by designers.
3. Green material management, green construction, and green operation maintenance phases. In these phases, contractors and material suppliers were invited as the respondents.

The details of the respondents involved in the indicator suggestion process are listed in Table 1. This sample seemed to satisfy the minimum number of respondents, according to Okoli and Pawlowski [56].

Table 1. Details of the respondents.

Respondent	Role	Job Title/Field of Expertise	Experience (Years)
1.	Academic	Civil engineering	>25
2.	Academic	Environmental engineering	>25
3.	Practitioner	Civil engineer	>25
4.	Practitioner	General manager	>25
5.	Practitioner	Engineer	>25
6.	Practitioner	Head of operation division	20
7.	Academic	Architectural engineering	>25
8.	Practitioner	Assistant manager of engineering and quality	>25
9.	Practitioner	Procurement engineer	5
10.	Practitioner	Supervisor project	>25
11.	Practitioner	Production officer	>25
12.	Practitioner	Project manager	>25
13.	Practitioner	Engineering and standardization officer	4
14.	Practitioner	Building information modeling (BIM) expert	>25
15.	Practitioner	Knowledge management officer	>25
16.	Practitioner	Director of human capital management and system development	>25

Second, the final evaluation model for RL performance was redistributed to the academics and practitioners in the form of questionnaire to allow them to assess the indicators. This questionnaire used a Likert scale to measure the relevance of certain indicators to measuring RL performance. Purposive sampling was also used in this research.

Finally, after all the data from respondents were collected, content validity analysis was carried out by calculating the content validity ratio (*CVR*). *CVR* is a numeric value that indicates the instrument's degree of validity determined by the experts' ratings of content validity. The sequence of steps to validate constructs and indicators using the content validity index is as follows [57]:

- Step 1: Determine the rating scale to be used to validate the constructs, concepts, elements, and indicators. The rating is 1 if the indicator is not relevant, 2 if the indicator is quite relevant, and 3 if the indicator is highly relevant.
- Step 2: Send the questionnaire to the respondents. The minimum number of respondents used to validate the results of the performance measurement indicators is at least ten [56].
- Step 3: Based on the returned responses, calculate the value of the *CVR*, which is a calculation method that linearly transforms the proportion of respondents who agree to the construct, concept, element, and indicator being tested. The formula for calculating the *CVR* can be seen in Equation (1) [58].

$$CVR = \frac{n_e - \left(\frac{N}{2}\right)}{\frac{N}{2}} \tag{1}$$

where

CVR: content validity ratio
n_e: the number of experts who gave a rating of 3 or relevant
N: the number of all experts

- Step 4: Eliminate irrelevant constructs, concepts, elements, and indicators.

3.1. Desk-Based Research

The desk-based research was conducted to collate the indicators used to measure RL performance in the construction sector. The research was performed by searching for previous studies on Scopus and the Web of Science using a combination of keywords such as "reverse logistics", "reverse logistics performance assessment", "reverse supply chain", "reverse logistics construction sector" and "waste management construction sector". Figure 2 shows the collation of RL practices and the generation of the RL performance via RL practices that are influenced by the initiation phase in the PLC, in terms of drivers and barriers.

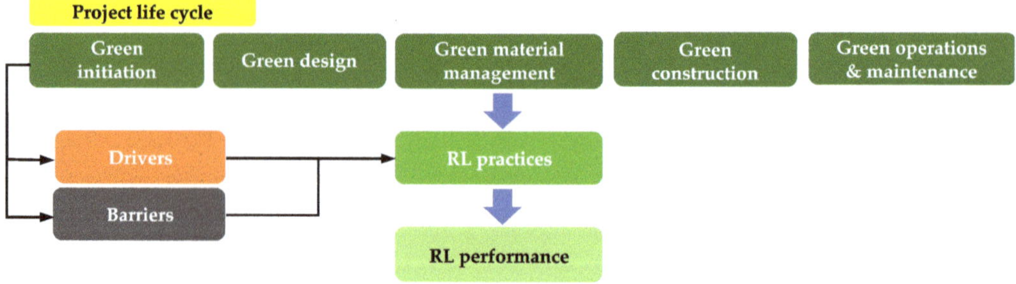

Figure 2. The relationship between PLC, RL practice, and RL performance.

Previous studies in building projects that have used the proposed framework include Wibowo et. al., Pushpamali et al., Hammes et al., Farida et al. [10,12,15,17]. Hammes et al. [15] suggested an assessment of RL performance in terms of the activities carried out in the construction phase. Wibowo et al. [10] focused on developing the concept of GSCM based on a PLC. Their study resulted in the five following basic concepts of GSCM application in the construction sector: (1) green initiation, (2) green design, (3) green material management, (4) green construction, and (5) green O/M.

Farida et al. [17] developed a GSCM assessment model for the construction sector. Pushpamali et al. [12] demonstrated that RL is strongly influenced by the decision to implement RL in the preconstruction phase or during project initiation, so the measuring tool developed should assess RL implementation from the initial phase, specifically from green initiation to the final phase of the project. Furthermore, regarding the proposed framework, a literature study related to indicators of RL performance measurement was carried out based on the PLC.

Table 2 lists the 66 indicators collected from the green initiation, green design, green material management, green construction, and green O/M phases.

Table 2. Initial RL measurement indicators in the construction sector collated from the desk-based research.

No	Element	Indicator	Code	References
		Green Initiation Phase		
		Dimension: Commitment		
1.	General commitment	Managerial resource	RC1	[54,59]
		Selection criteria	RC2	[60]
2.	Resource efficient commitment	Recycled content	RC3	[60]
		Materials transportation	RC4	[60]
		Technical specification: low temperature asphalt	RC5	[60]
		Soil and waste management plan	RC6	[60]
		Dimension: Feasibility study		
3.	Economic assessment	Saving in material cost	FS1	[61]
		Reduction in waste	FS2	[61]
		Life cycle cost	FS3	[61]
4.	Customer perceived level of service	Percentage of customer willingness	FS4	[62]
		Dimension: Knowledge management process		
5.	Knowledge application process	Problem sharing	KM1	[63]
		Best practice sharing	KM2	[63]
		Green Design Phase		
		Dimension: Design innovation		
6.	Material efficiency	Material efficiency index	IDI1	[64]
		Reusable or recyclable material	IDI2	[64]
		Dimension: Knowledge management process		
7.	Knowledge application process	Design change improvement	KM1	[64]
		Dimension: Guideline for deconstruction design		
8.	Deconstruction design (DFD) for recycle material	Using recycled materials	GD1	[63]
		Avoiding use of hazardous and toxic materials	GD2	[65]
		Green Material Management Phase		
		Dimension: Green purchasing practices		
9.	Green supplier selection	Cost: raw material price	SSC1	[66]
		Cost: product	SSC2	[67]
		Cost: logistics	SSC3	[67,68]
		Reject rate	SSQ1	[67]
		Delivery capabilities	SSD1	[67,68]

Table 2. *Cont.*

No	Element	Indicator	Code	References
		Green Material Management Phase		
		Dimension: Green purchasing practices		
		Order fulfilment rate	SSD2	[67]
		Production capacity	SSD3	[67]
		Energy consumption	SSE1	[67,68]
		Wastewater treatment	SSE2	[67]
		Environmental staff training	SSE3	[67]
		Environmentally friendly material	SSE4	[68]
		Environmentally friendly planning	SSE5	[68]
		Capability of deconstruction/ disassembly design	SSI6	[67,68]
		Speed of development	SSI7	[67,68]
		Safety assurance	SSS1	[67]
10.	Supplier safety performance	Loss time accident (LTA)	SSS2	[69]
		Occupational Health and Safety (OHS)	SSS3	[69]
		Personal protective equipment (PPE)	SSS4	[68]
		Expert certification OHS	SSS5	[69]
		Safety induction	SSS6	[69]
		OHS policy	SSS7	[69]
11.	Green supplier development	Quality evaluation	SDQ1	[66]
		Delivery evaluation	SDD1	[66]
12.	Green supplier collaboration	Supplier risk assurance	SCC1	[67]
		Safety assurance	SCC2	[67]
13.	Green supplier evaluation	Quality evaluation	SEQ1	[66]
		Delivery evaluation	SED1	[66]
		Cost evaluation	SEC1	[66]
		Green Construction Phase		
		Dimension: Knowledge process management		
14.	Design change	Design change implementation	KM3	[63]
		Dimension: RL Practices		
15.	RL supplier side	Green purchase	RLSS1	[15]
16.	RL internal side	Use of material	RLIS1	[15]
		Reuse of material	RLIS2	[15]
		Recycling material	RLIS3	[17]
		Remanufacture	RLIS4	[17]
		Residual	RLIS5	[17]
		Return on investment (ROI)	RLIS6	[15]
		Customer satisfaction	RLIS7	[15]
17.	RL waste management side	Storage 1	RLWM1	[15]
		Storage 2	RLWM2	[15]
		Transportation 1	RLWM3	[15]
		Transportation 2	RLWM4	[15]
		Transportation 3	RLWM5	[15]
		Worker awareness 1	RLWM6	[15]
		Worker awareness 2	RLWM7	[15]
		Dimension: Safety		
18.	Safety	Safety performance	SF1	[60]
		Green Operations and Maintenance (O/M) Phase		
		Dimension: Waste management plan		
19.	Waste management plan	Technical specification: tar containing asphalt	WMP1	[60]
		Dimension: Durability		
20.	Durability	Service lifetime	DR1	[60]
		Dimension: Safety		
21.	Safety	Safety performance	SF1	[60]
		Dimension: Knowledge sharing management		
22.	Knowledge application process	Problem sharing	KM1	[63]

3.2. FGD

In this step, the 16 respondents were given closed questionnaires asking whether or not the indicators in Table 2 are relevant in measuring RL performance in the construction sector. The respondents stated that the 66 indicators can be considered as tools for measuring RL performance in the construction sector. The respondents were then asked whether there were additional indicators for measuring RL performance in each phase of the PLC. As a result, nine additional indicators were proposed by the respondents, as shown in Table 3.

Table 3. Additional indicators proposed by academics and practitioners.

No.	Code	Indicator	Definition	Phase	Dimension
1.	ISR8	Total RL principles applied in the project	RL principles stated in the project agreement, such as requests to reuse and recycle materials	Green Initiation	Commitment
2.	MSR8	RL clause in the instruction to bidder	Existence of a clause that regulates the supplier's obligation to carry out RL	Green Material Management	Green Procurement Practices
3.	MSR9	Preparation of material priority scale plan for RL implementation	Existence of a plan to develop a material priority scale in an effort to implement RL	Green Material Management	Green Procurement Practices
4.	CSR16	Domestic content level	Percentage of the material content of domestic/local products in the whole project	Green Construction	RL Practices
5.	CSR8	Evaluation of quality, cost, and time in the results	Evaluation of quality, cost, and time on the results of construction projects that apply RL	Green Construction	RL Practices
6.	OSR6	O/M energy usage	Consumption of all the energy used to perform an action, manufacture an item, or simply inhabit a building	Green O/M	Knowledge Sharing Management
7.	OSR16	Percentage of repairs in O/M phase due to material damage	A number indicating the reliability of a system/equipment based on a review of repair costs over a period of time	Green O/M	Knowledge Sharing Management
8.	OSR61	Capacity factor	The ratio of the total actual energy produced or supplied over a definite period to the energy that would have been produced if the plant (generating unit) had operated continuously at the maximum rating	Green O/M	Knowledge Sharing Management
9.	OSR8	Corrective and preventive actions	Existence of corrective and preventive actions if there is a problem related to the implementation of RL during the maintenance process	Green O/M	Knowledge Sharing Management

3.3. Validation of the RL Measurement Indicators

To eliminate items that do not represent relevant measures to be carried out, the results of the *CVR* calculation were compared with the *CVR* minimum value guideline table based on the number of experts by Lawshe [58]. The minimum value of the *CVR* with 16 experts is 0.5.

The indicators generated from the literature study and FGD were compiled based on the PLC phases in GSCM. The reason for compiling a list of RL indicators based on the PLC is to incorporate RL from the beginning of the construction process, namely initiation and design construction. There are 75 indicators for various PLC phases in GSCM. The indicators list was distributed to respondents to provide scores related to the suitability of indicators in each phase and to add indicators based on best practices and respondents' experiences. The results of the assessment were analyzed using the *CVR*, as shown in Table 4.

Table 4. Results of the RL measurement validation.

No.	Element	Indicator	Code	References
		Green Initiation Phase		
		Dimension: Commitment		
1.	General commitment	Managerial resource	RC1	[54]
		Selection criteria	RC2	[60]
2.	Resource efficient commitment	Total RL principles applied in the project	ISR8	FGD
		Dimension: Feasibility study		
3.	Economic assessment	Saving in material cost	FS1	[61]
		Reduction in waste	FS2	[61]
		Dimension: Knowledge management process		
4.	Knowledge application process	Best practice sharing	KM2	[63]
		Green Design Phase		
		Dimension: Design innovation		
5.	Material efficiency	Material efficiency index	IDI1	[64]
		Reusable or recyclable material	IDI2	[64]
		Dimension: Guideline for deconstruction design		
6.	DFD for recycled material	Using recycled materials	GD1	[63]
		Avoiding use of hazardous and toxic materials	GD2	[65]
		Green Material Management Phase		
		Dimension: Green purchasing practices		
7.	Green supplier selection	Cost: Raw material price	SSC1	[66]
		Cost: Product	SSC2	[67]
		Dimension: Green procurement practices		
8.	Green procurement practices	RL clause in the instruction to bidder	MSR8	FGD
		Preparation of material priority scale plan for RL implementation	MSR9	FGD
		Green Construction Phase		
		Dimension: RL practices		
9.	RL internal side	Use of material	RLIS1	[15]
		Reuse of material	RLIS2	[15,17]
		Recycling material	RLIS3	[17]
		Residual	RLIS5	[17]
10.	RL waste management side	Evaluation of quality, cost, and time on the results of construction projects that apply RL	CSR8	FGD
11.	Knowledge application process	Problem sharing	KM1	[63]
		Percentage of repairs in O/M phase due to material damage	OSR16	FGD

4. Synthesis of the RL Performance Measurement for the Construction Industry

An RL performance evaluation indicator that integrates each phase in the PLC needs to be developed as a first step to determine the performance of the construction sector in implementing RL. In this study, the RL evaluation indicator was developed by adopting RL performance indicators from the manufacturing sector. Based on the results of the *CVR*, as shown in Table 4, 21 indicators have a *CVR* value greater than the minimum *CVR* value (>0.5). Therefore, 21 indicators are considered valid for measuring RL performance in the construction sector, as shown in Figure 3.

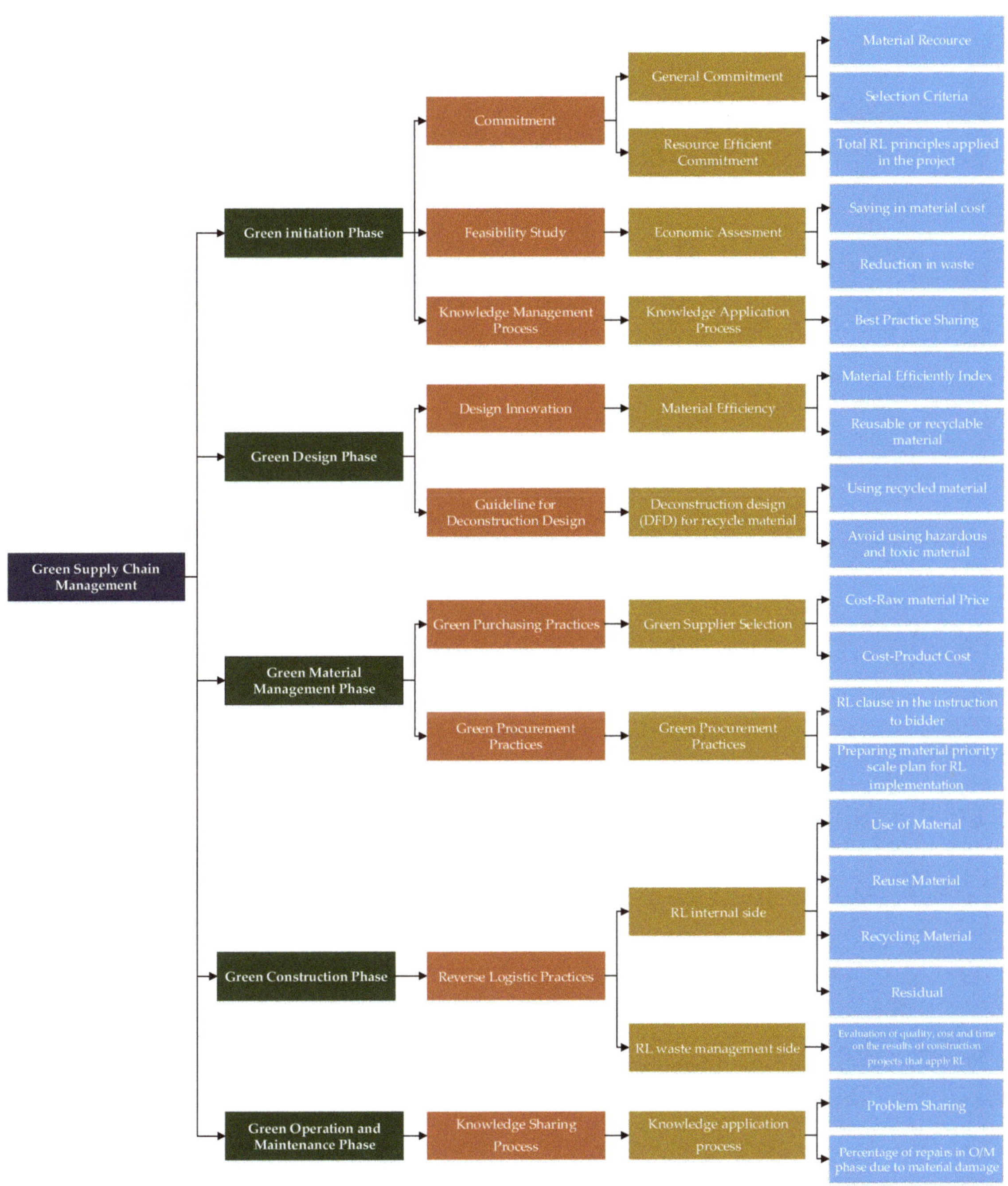

Figure 3. The final RL performance measurement for the construction industry.

In the construction industry, the duration of projects is typically long, and the phases (initiation, design, material management, construction, operation, and maintenance) are integrated. The present study differs from that of Hammes et al. [15], which measured RL performance based on supplies, internal logistics, and waste management. While their research focused only on the construction phase, using 12 RL measurement indicators, this study develops the concept of RL measurement on the basis of the PLC with 21 measurement indicators. The present research is more robust because the concept of PLC [10], as the basis for measuring RL, continues through several stages, namely the desk-based research and the validation process carried out by 16 people in the construction industry. This study also develops RL measurements based on the research of Wibowo et al. [10] and Pushpamali et al. [12], whose formation of RL indicators involves three stages, namely, desk-based research, FGD and validation involving the experts. However, the present research is also more robust than these studies because the measurements are carried out at each phase in the PLC, namely green initiation, green design, green material management, green construction, and green O/M.

The green initiation and design phases play important roles in supporting RL performance measurement. In the initiation phase, the stakeholder (owner) must ensure that the project being built is sustainable, taking into consideration the work of the architect in the design phase. In the design phase, the DED implementation should consider the guideline for deconstruction design. The green material management phase also involves using eco-friendly materials to replace non-eco-friendly materials according to the previous phase. The green construction phase can incur an enormous amount of waste, but if the project already uses eco-friendly material, both waste and emissions will be reduced. When the reuse and recycling of material is successfully applied according to the project conditions in the field, the implementation of RL becomes easier, and so as controlling energy consumption becomes more efficient in the green O/M phase. Therefore, the RL performance measurements need to be integrated throughout the PLC system.

In manufacturing companies, where the RL process takes place in one organization, one location, or one work unit (blended), it is relatively easy to apply and control SCM related to material, information, and financial flows. In contrast, in companies operating in the construction sector, each stakeholder involved in measuring RL performance may work with different organizations (consisting of three or more organizations) or fragmented project owners, contractors, and consultant teams within a certain period. The role of stakeholders, especially in construction projects, is very important.

Previous research has emphasized that the stakeholders in the construction sector can be a decisive factor in "making or breaking" a project [70]. Therefore, the commitment of stakeholders to construction projects is important because they come from different organizations, educational backgrounds, and specializations to perform a task within certain time limits and with certain goals. Thus, it is necessary to establish a common premise of shared interest in the building project. If stakeholders in each phase do not have the same rationale, values, or spirit, RL will be difficult to implement. Therefore, the importance of the PLC approach is in its ability to unite or link the understanding and values of stakeholders on the basis of RL.

4.1. The Roles of Project Owners in the Green Initiation Phase

Green initiation is the initial stage in the implementation of a project. In this phase, the value or spirit of the project requirements is an important aspect in implementing GSCM. Establishing this value helps create collaboration between stakeholders in a project, allowing the project's goals to be achieved [71].

In the green initiation phase, there are six indicators that are considered valid in this study, comprising (1) RC1: managerial resource, (2) RC2: selection criteria, (3) FS1: saving material cost, (4) FS2: reduction in waste, (5) KM2: best practice sharing about green projects, and (6) ISR8: total RL principles applied in the project. The green initiation phase is related to the project owner's commitment to implementing green aspects in the project to be made. The project owner is a key stakeholder because they have the authority to decide project criteria and also commitment is the main determining factor in implementing environmentally friendly projects. These results are in line with the indicators proposed by Olanipekun et al. [72], who found that, from the perspective of various stakeholders, the project owner's commitment depended on their experience and capability in handling green building projects. Having workforce who are capable of applying various aspects of the project that have environmental impact is also crucial to the implementation of RL in terms of managerial resources (RC1) [54]. Furthermore, FS1 (saving material cost) and FS2 (reduction in waste) are indicators used to measure the feasibility dimension when evaluating RL performance on a green project. Saving material cost (FS1) shows the estimated profit obtained if the project uses recycled materials and the estimated waste (FS2) that can be derived from the use of recycled materials. These two indicators to measure the feasibility of RL implementation are adapted from research by Halil et al. [61], where both indicators are used to assess the feasibility of implementing green construction from an economic perspective [61]. Research by Tan et al. [73] has shown that the economic aspect is the main consideration in determining the feasibility of a green project as the results of such feasibility studies influence the owner's decisions in setting project criteria, such as RL implementation.

4.2. Material Efficiency Index in the Construction Design Phase

Design is defined as the process of developing a solution to a particular problem using the necessary experts and tools. It is a step in the planning process where a detailed description is produced that reflects the project concept. Importantly, green parameters and sustainable construction occur only when the environmental, social, and economic considerations are addressed and incorporated into the design process [74]. In sustainable design, social, environmental, and economic factors need to be taken into consideration before designing any construction project. Studies should be conducted regarding the ability to supply raw materials and whether the building users benefit from using minimum resources with less damage to the environment [75]. A well-defined design policy among stakeholders can also be crucial before starting a project with a green project concept. Some researchers believe that designers can make changes to the design mentality and process to engage in green issues.

The construction requirements for any sustainable project should be decided on prior to the construction phase, and sufficient time should be spent to come up with an appropriate plan to avoid changes during construction and to save time and cost [75]. Therefore, designers must be involved in the project process from the initial stage—the "planning stage"—to incorporate effective changes related to the green project concept [74].

The indicators in the green design phase declared valid in this study are the material efficiency index (ID1), recycling material (ID2), use of recycled materials (GD1), and the level of use of hazardous materials (GD2). By using the material efficiency indicator (ID1), the company adopts a system capable of tracking the use of all materials from the beginning of processing until the material reaches the end of its useful life. Hence, with material efficiency as an indicator, the company controls how a material is reused, recycled, and remanufactured. Controlling the use and selection of materials is a means for companies to determine ideas for improvement, one of which is through the implementation of RL as this improvement aims to increase the efficiency of the material index. Furthermore, with the ID2 (reusable or recyclable material) indicator, designers become more conscious of making designs that are environmentally friendly and easy to disassemble. Through the application of environmentally friendly design concepts, waste problems caused by the construction process can be overcome. The application of environmentally friendly design concepts also facilitates the implementation of RL [76]. The use of recycled material (GD1) is one indicator used to measure eco-design. Its aim is to reduce the use of virgin materials so that the availability of materials can be maintained in the long term [59], and the company can obtain cost savings by purchasing recycled materials in procurement activities. The use of nontoxic or nonhazardous materials (GD2) helps ensure that the deconstruction results from implementing RL do not endanger workers when used [65].

4.3. Green Procurement in the Green Material Management Phase

Material management is the system for planning and controlling to ensure that the correct quality and quantity of materials and equipment are specified in a timely manner. Materials should be obtained at a reasonable cost and be available for use when needed. The cost of materials represents a large proportion of the overall construction cost. Therefore, the role of stakeholders in controlling the management of RL in the green material management phase is essential because it provides the basis for the green construction phase related to field implementation. In the construction sector, RL performance in the green material management phase is also related to green procurement practices. Green procurement practices include green supplier selection, supplier safety performance, green supplier development, green supplier collaboration, and green supplier evaluation activities. Green procurement practice is an important criterion in creating sustainability and plays a role in maintaining environmental performance to minimize impacts throughout the construction process [77].

The green material management indicators in this study are raw material price (SSC1) and product cost (SSC2). These two indicators are used to measure the performance of RL because both measure the profits obtained by the company when implementing RL. The indicators' relevance is reinforced by research by Škapa and Klapalová [78] regarding company profits. These two indicators are also able to measure the use of material resources in procurement activities. Resource use is the main indicator in the criteria for green public procurement projects for road construction. In addition, the indicators that are declared valid within the green procurement practices dimension are the MSR8 indicator (existence of an RL clause on the employee requirement/instruction to bidder) and the MSR9 indicator (existence of a plan to develop a material priority scale for RL implementation). According to previous research, the presence of the MSR8 and MSR9 indicators will guarantee the implementation of RL [79].

4.4. Reuse and Recycle in the Green Construction Phase

Green construction, as the next concept to engage with in the construction process of environmentally friendly buildings, is developed by various stakeholders. A particularly important stakeholder in this phase is the contractor. The contractor is tasked with planning, implementing, and supervising construction activities from start to finish to ensure that all aspects are in accordance with existing regulations. In this PLC concept, contractors are not only responsible for constructing strong and efficient buildings but must also pay attention to the environment. Green construction is an important phase in minimizing the environmental impact caused. The green construction approach seeks to balance the capabilities of the environment with the needs of human life for present and future generations [17] through the efficient use of resources [80]. The three main stages in green construction are reducing the use of non-environmentally friendly resources, reducing the waste generated during the process, and reducing the emissions generated by the project. The purpose of implementing green construction is to minimize waste at the construction stage indirectly by reducing energy and resources; as a result, emissions will also be reduced during the construction process [10].

However, there are several obstacles that prevent companies from implementing green construction. These include the following: (a) contractors being constrained by the limited availability of environmentally friendly equipment; (b) the unavailability of workers trained in the principles of green construction; (c) a lack of certainty about the type of environmentally friendly material declared by a legitimized institution; (d) technology limitations in implementing green construction; (e) no effective internal collaboration between large contractors and specialist contractors and (f) limited regulations governing green construction.

In the green construction phase, the relevant indicators are the use of material (RLIS1), reuse of material (RLIS2), recycling of material (RLIS3), residual of material (RLIS 5), and evaluation of quality, cost, and time. Some construction project results have applied RL (CSR8) as an indicator used to measure the performance of RL in the green construction phase. This indicator relates to operational activities during the construction process. Reuse and recycling are values that are measured in the application of RL, making these two indicators very important in measuring RL performance. Reuse and recycling in RL are also supported by Ripanti and Tjahjono [13].

4.5. Problem Sharing in Green O/M Phase

The green O/M phase is related to energy consumption as the largest energy consumption occurs in this phase from the perspective of life cycle costs [81]; thus, the implementation of RL performance measurement is critical. The green O/M phase involves project residents or users. Therefore, every stakeholder, especially the owner and building manager, needs a coordinated understanding of the importance of focusing on the occupants of the building. The indicators used to measure the performance of RL in the green O/M phase are problem sharing (KM1) and the percentage of repairs in the O/M phase due to material damage (OSR16). KM1 indicator is used to measure the RL application constraints that arise at the end of the phase so that the obstacles that arise can be anticipated from the beginning of the project. The percentage of repairs in the O/M phase due to material damage (OSR16) aims to determine the performance of the RL material used in the project. This indicator is in line with Abraham et al. [81], who state that an enterprise's preference in the O/M phase no longer requires significant investment. Creating a noticeably effective product from recycled aggregates makes the construction material substantially greener and more sustainable. These results can be assisted by the coordination of project managers and governing bodies in lowering the cost of the life cycle of materials that can be used in the homes. Through life cycle cost analysis, building owners can obtain detailed information about material costs, and the environmental impacts due to C&DW can be reduced by using the waste from other products.

5. Conclusions

RL is considered a remedial measure that moderates the detrimental impacts of construction projects on the natural environment and enables organizations to be more efficient and effective by attaining economic benefits and sustainable competitiveness. In this case, RL aims to increase the value of waste generated by construction activities and reduce costs for waste management. To gain a better perception of the RL of companies, RL performance measurement should be implemented throughout the whole PLC.

5.1. Theoretical Contributions

Performance evaluation of RL practices in construction sectors is crucial, but only few studies have focused on measuring the RL performance. Most research on RL in the construction sector, e.g., Pushpamali et al. [12], has not specifically provided an evaluation of the RL performance and, in this respect, they seemed to focus only on the construction phase, e.g., Hammes et al. [15] and Farida et al. [17].

This paper contributes to the construction sector's literature by presenting a new, PLC-based perspective on how RL can be adopted, from the initiation, design, materials management, and construction phases to the O/M phases. It also enhances the research area of Green Supply Chain Management (GSCM) in construction, where the RL performance has become an important factor for the construction sector in order to be more environmentally conscious.

Finally, the paper proposes a new model that integrates the work of Wibowo et al. [10], Pushpamali et al. [12], Hammes et al. [15], and Farida et al. [17]. The model consists of dimensions, elements, and indicators for the evaluation of RL performance throughout the construction's PLC.

5.2. Implications for Practice

This paper identifies the Green Supply Chain Management (GSCM) as an important platform that enables the stakeholders to get involved in each phase of the PLC. The scope of PLC includes the initiation and design phases, two critical phases that determine the success of RL. The development of the RL measurement model starts with the construction of each phase in the PLC. In the initiation phase, the building owners play a vital role in creating/building environmentally friendly value by applying RL to the constructions. The RL concept should also be realized in the DED made by the design consultant, creating the DFD.

The environmentally friendly results of the construction project are then handed back to the owners, who continue applying environmentally friendly values during the O/M phase. The role of each stakeholder during the construction process of a building or infrastructure is critical. The environmentally friendly value based on the PLC and in accordance with GSCM and RL applications must be implemented by all stakeholders.

5.3. Limitations and Future Work

This paper has some limitations. First, the selection of the participants of FGD, though involving a wide range of stakeholders who were truly independent experts at every phase of the PLC, was based on a purposive sampling. This, arguably, relied on the personal opinion of the participants. Second, the use of questionnaire to validate the measurements by a relatively small number of respondents might lead to bias though this has been mitigated by closely liaising with them and, at the same time, ensuring their responses were kept anonymous and confidential.

With respect to the abovementioned limitations, the performance measures of RL practices in the construction sector proposed in this paper are thus considerably conceptual in nature. Future research should therefore look into applying the measures to real building projects, in order to ascertain their practical relevance.

Author Contributions: Conceptualization, M.A.W., N.U.H. and B.T.; methodology, N.U.H.; formal analysis, M.A.W. and A.M.; investigation, S.A.W.; writing—original draft preparation, M.A.W., S.A.W. and A.M.; writing—review and editing, B.T.; supervision, M.A.W. and B.T. All authors have read and agreed to the published version of the manuscript.

Funding: This research was funded by the Institute for Research and Community Services—Diponegoro University under the research grant of World Class Research University, grant number 118-29/UN7.6.1/PP/2021.

Institutional Review Board Statement: Not applicable.

Informed Consent Statement: Not applicable.

Data Availability Statement: Not applicable.

Acknowledgments: The authors are grateful for the funding support from the Institute for Research and Community Services—Diponegoro University, under the research grant of World Class Research University, which enabled the bilateral research collaboration between the Faculty of Engineering, Diponegoro University and the Centre for Business in Society, Coventry University.

Conflicts of Interest: The authors declare no conflict of interest.

References

1. Sobotka, A.; Sagan, J.; Baranowska, M.; Mazur, E. Management of reverse logistics supply chains in construction projects. *Procedia Eng.* **2017**, *208*, 151–159. [CrossRef]
2. Wibowo, M.A.; Handayani, N.U.; Nurdiana, A.; Sholeh, M.N.; Pamungkas, G.S. Mapping of information and identification of construction waste at project life cycle. *AIP Conf. Proc.* **2018**, *1941*, 020049. [CrossRef]
3. United Nations Environment Programme. *Buildings and Climate Change*; UNEP DTIE Sustainable Consumption and Production Branch: Paris, France, 2009.

4. Surahman, U.; Higashi, O.; Kubota, T. Evaluation of current material stock and future demolition waste for urban residential buildings in Jakarta and Bandung, Indonesia: Embodied energy and CO_2 emission analysis. *J. Mater. Cycles Waste Manag.* **2017**, *19*, 657–675. [CrossRef]
5. Cheng, J.C.P.; Law, K.H.; Bjornsson, H.; Jones, A.; Sriram, R. A service oriented framework for construction supply chain integration. *Autom. Constr.* **2010**, *19*, 245–260. [CrossRef]
6. Shin, T.-H.; Chin, S.; Yoon, S.-W.; Kwon, S.-W. A service-oriented integrated information framework for RFID/WSN-based intelligent construction supply chain management. *Autom. Constr.* **2011**, *20*, 706–715. [CrossRef]
7. Chileshe, N.; Rameezdeen, R.; Hosseini, M.R.; Lehmann, S.; Udeaja, C. Analysis of reverse logistics implementation practices by South Australian construction organisations. *Int. J. Oper. Prod. Manag.* **2016**, *36*, 332–356. [CrossRef]
8. Ghobakhloo, M.; Tang, S.H.; Zulkifli, N.; Ariffin, M. An integrated framework of green supply chain management implementation. *Int. J. Innov. Manag. Technol.* **2013**, *4*, 86.
9. Granlie, M.; Hvolby, H.H.; Cassel, R.A.; Paula, I.C.D.; Soosay, C. A Taxonomy of Current Literature on Reverse Logistics. *IFAC Proc. Vol.* **2013**, *46*, 275–280. [CrossRef]
10. Wibowo, M.A.; Handayani, N.U.; Mustikasari, A. Factors for implementing green supply chain management in the construction industry. *J. Ind. Eng. Manag.* **2018**, *11*, 651–679. [CrossRef]
11. Agrawal, S.; Singh, R.K.; Murtaza, Q. A literature review and perspectives in reverse logistics. *Resour. Conserv. Recycl.* **2015**, *97*, 76–92. [CrossRef]
12. Pushpamali, N.N.C.; Agdas, D.; Rose, T.M. A Review of Reverse Logistics: An Upstream Construction Supply Chain Perspective. *Sustainability* **2019**, *11*, 4143. [CrossRef]
13. Ripanti, E.F.; Tjahjono, B. Unveiling the potentials of circular economy values in logistics and supply chain management. *Int. J. Logist. Manag.* **2019**, *30*, 723–742. [CrossRef]
14. Chinda, T. Examination of Factors Influencing the Successful Implementation of Reverse Logistics in the Construction Industry: Pilot Study. *Procedia Eng.* **2017**, *182*, 99–105. [CrossRef]
15. Hammes, G.; De Souza, E.D.; Taboada Rodriguez, C.M.; Rojas Millan, R.H.; Mojica Herazo, J.C. Evaluation of the reverse logistics performance in civil construction. *J. Clean. Prod.* **2020**, *248*, 119212. [CrossRef]
16. Hosseini, M.R.; Rameezdeen, R.; Chileshe, N.; Lehmann, S. Reverse logistics in the construction industry. *Waste Manag. Res.* **2015**, *33*, 499–514. [CrossRef]
17. Farida, N.; Handayani, N.U.; Wibowo, M.A. Developing Indicators of Green Construction of Green Supply Chain Management in Construction Industry: A Literature Review. *IOP Conf. Ser. Mater. Sci. Eng.* **2019**, *598*, 012021. [CrossRef]
18. Bernon, M.; Tjahjono, B.; Ripanti, E.F. Aligning retail reverse logistics practice with circular economy values: An exploratory framework. *Prod. Plan. Control* **2018**, *29*, 483–497. [CrossRef]
19. Tjahjono, B.; Fernandez, R. Practical approach to experimentation in a simulation study. In Proceedings of the 2008 Winter Simulation Conference, 7–10 December 2008; pp. 1981–1988.
20. Murphy, P. A Preliminary Study of Transportation and Warehousing Aspects of Reverse Distribution. *Transp. J.* **1986**, *25*, 12–21.
21. Lambert, D.M.; Stock, J.R. *Strategic Physical Distribution Management*; RD Irwin: Homewood, IL, USA, 1982.
22. Murphy, P.R.; Poist, R.F. *Management of Logistical Retromovements: An Empirical Analysis of Literature Suggestions*; Transportation Research Forum: Washington, DC, USA, 1988.
23. Prajapati, H.; Kant, R.; Shankar, R. Bequeath life to death: State-of-art review on reverse logistics. *J. Clean. Prod.* **2019**, *211*, 503–520. [CrossRef]
24. Guarnieri, P.; Camara e Silva, L.; Vieira, B.D. How to Assess Reverse Logistics of e-Waste Considering a Multicriteria Perspective? A Model Proposition. *Logistics* **2020**, *4*, 25. [CrossRef]
25. Rogers, D.S.; Tibben-Lembke, R.S. *Going Backwards: Reverse Logistics Trends and Practices. The University of Nevada, Reno*; Center for Logistics Management, Reverse Logistics Council: Reno, NV, USA, 1998; pp. 2–33.
26. de Oliveira, U.R.; Aparecida Neto, L.; Abreu, P.A.F.; Fernandes, V.A. Risk management applied to the reverse logistics of solid waste. *J. Clean. Prod.* **2021**, *296*, 126517. [CrossRef]
27. Guarnieri, P.; e Silva, L.C.; Levino, N.A. Analysis of electronic waste reverse logistics decisions using Strategic Options Development Analysis methodology: A Brazilian case. *J. Clean. Prod.* **2016**, *133*, 1105–1117. [CrossRef]
28. Accorsi, R.; Manzini, R.; Pini, C.; Penazzi, S. On the design of closed-loop networks for product life cycle management: Economic, environmental and geography considerations. *J. Transp. Geogr.* **2015**, *48*, 121–134. [CrossRef]
29. Guide, V.D.R.; Van Wassenhove, L.N. OR FORUM—The Evolution of Closed-Loop Supply Chain Research. *Oper. Res.* **2009**, *57*, 10–18. [CrossRef]
30. Augiseau, V.; Barles, S. Studying construction materials flows and stock: A review. *Resour. Conserv. Recycl.* **2017**, *123*, 153–164. [CrossRef]
31. Eckelman, M.J.; Brown, C.; Troup, L.N.; Wang, L.; Webster, M.D.; Hajjar, J.F. Life cycle energy and environmental benefits of novel design-for-deconstruction structural systems in steel buildings. *Build. Environ.* **2018**, *143*, 421–430. [CrossRef]
32. Shen, L.Y.; Tam Vivian, W.Y.; Tam, C.M.; Drew, D. Mapping Approach for Examining Waste Management on Construction Sites. *J. Constr. Eng. Manag.* **2004**, *130*, 472–481. [CrossRef]
33. Arora, M.; Raspall, F.; Cheah, L.; Silva, A. Buildings and the circular economy: Estimating urban mining, recovery and reuse potential of building components. *Resour. Conserv. Recycl.* **2020**, *154*, 104581. [CrossRef]

34. Hosseini, M.; Chileshe, N.; Rameezdeen, R.; Lehmann, S. *Sensitizing the Concept of Reverse Logistics (RL) for Construction Context*; Islamic Azad University: Tehran, Iran, 2013.
35. Esin, T.; Cosgun, N. A study conducted to reduce construction waste generation in Turkey. *Build. Environ.* **2007**, *42*, 1667–1674. [CrossRef]
36. Dowlatshahi, S. A strategic framework for the design and implementation of remanufacturing operations in reverse logistics. *Int. J. Prod. Res.* **2005**, *43*, 3455–3480. [CrossRef]
37. Shakantu, W.; Muya, M.; Tookey, J.; Bowen, P. Flow modelling of construction site materials and waste logistics. *Eng. Constr. Archit. Manag.* **2008**, *15*, 423–439. [CrossRef]
38. da Rocha, C.G.; Sattler, M.A. A discussion on the reuse of building components in Brazil: An analysis of major social, economical and legal factors. *Resour. Conserv. Recycl.* **2009**, *54*, 104–112. [CrossRef]
39. Sinha, S.; Shankar, R.; Taneerananon, P. Modelling and case study of reverse logistics for construction aggregates. *Int. J. Logist. Syst. Manag.* **2009**, *6*, 39–59. [CrossRef]
40. Sassi, P. Defining closed-loop material cycle construction. *Build. Res. Inf.* **2008**, *36*, 509–519. [CrossRef]
41. Vandecasteele, C.; Heynen, J.; Goumans, H. Materials Recycling in Construction: A Review of the Last 2 Decades Illustrated by the WASCON Conferences. *Waste Biomass Valorization* **2013**, *4*, 695–701. [CrossRef]
42. Ghaffar, S.H.; Burman, M.; Braimah, N. Pathways to circular construction: An integrated management of construction and demolition waste for resource recovery. *J. Clean. Prod.* **2020**, *244*, 118710. [CrossRef]
43. Duan, H.; Li, J. Construction and demolition waste management: China's lessons. *Waste Manag. Res.* **2016**, *34*, 397–398. [CrossRef]
44. Huang, B.; Wang, X.; Kua, H.; Geng, Y.; Bleischwitz, R.; Ren, J. Construction and demolition waste management in China through the 3R principle. *Resour. Conserv. Recycl.* **2018**, *129*, 36–44. [CrossRef]
45. Yuan, H.; Shen, L.; Wang, J. Major obstacles to improving the performance of waste management in China's construction industry. *Facilities* **2011**, *29*, 224–242. [CrossRef]
46. Badenhorst, A. A framework for prioritising practices to overcome cost-related problems in reverse logistics: Original research. *J. Transp. Supply Chain Manag.* **2013**, *7*, 1–10. [CrossRef]
47. Micheli, P.; Mari, L. The theory and practice of performance measurement. *Manag. Account. Res.* **2014**, *25*, 147–156. [CrossRef]
48. Ying, F.; Tookey, J.; Seadon, J. Measuring the invisible. *Benchmarking Int. J.* **2018**, *25*, 1921–1934. [CrossRef]
49. Costa Dayana, B.; Formoso Carlos, T.; Kagioglou, M.; Alarcón Luis, F.; Caldas Carlos, H. Benchmarking Initiatives in the Construction Industry: Lessons Learned and Improvement Opportunities. *J. Manag. Eng.* **2006**, *22*, 158–167. [CrossRef]
50. Wegelius-Lehtonen, T. Performance measurement in construction logistics. *Int. J. Prod. Econ.* **2001**, *69*, 107–116. [CrossRef]
51. Shaik, M.; Abdul-Kader, W. Performance measurement of reverse logistics enterprise: A comprehensive and integrated approach. *Meas. Bus. Excell.* **2012**, *16*, 23–34. [CrossRef]
52. Bansia, M.; Varkey, J.K.; Agrawal, S. Development of a Reverse Logistics Performance Measurement System for a Battery Manufacturer. *Procedia Mater. Sci.* **2014**, *6*, 1419–1427. [CrossRef]
53. Guimarães, J.L.d.S.; Salomon, V.A.P. ANP Applied to the Evaluation of Performance Indicators of Reverse Logistics in Footwear Industry. *Procedia Comput. Sci.* **2015**, *55*, 139–148. [CrossRef]
54. Morgan, T.R.; Tokman, M.; Richey, R.G.; Defee, C. Resource commitment and sustainability: A reverse logistics performance process model. *Int. J. Phys. Distrib. Logist. Manag.* **2018**, *48*, 164–182. [CrossRef]
55. Etikan, I.; Musa, S.A.; Alkassim, R.S. Comparison of convenience sampling and purposive sampling. *Am. J. Theor. Appl. Stat.* **2016**, *5*, 1–4. [CrossRef]
56. Okoli, C.; Pawlowski, S.D. The Delphi method as a research tool: An example, design considerations and applications. *Inf. Manag.* **2004**, *42*, 15–29. [CrossRef]
57. Taherdoost, H. Validity and reliability of the research instrument; how to test the validation of a questionnaire/survey in a research. In How to Test the Validation of a Questionnaire/Survey in a Research (August 10, 2016). *Int. J. Acad. Res. Manag. (IJARM)* **2016**, *5*, 28–36.
58. Lawshe, C.H. A Quantitative Approach to Content Validity. *Pers. Psychol.* **1975**, *28*, 563–575. [CrossRef]
59. Khor, K.S.; Udin, Z.M. Reverse logistics in Malaysia: Investigating the effect of green product design and resource commitment. *Resour. Conserv. Recycl.* **2013**, *81*, 71–80. [CrossRef]
60. Garbarino, E.; Quintero, R.R.; Donatello, S.; Wolf, O. Revision of Green Public Procurement Criteria for Road Design, Construction and Maintenance. Procurement Practice Guidance Document. 2016. Available online: https://core.ac.uk/download/pdf/81684449.pdf (accessed on 20 July 2020).
61. Halil, F.M.; Nasir, N.M.; Hassan, A.A.; Shukur, A.S. Feasibility Study and Economic Assessment in Green Building Projects. *Procedia—Soc. Behav. Sci.* **2016**, *222*, 56–64. [CrossRef]
62. Lambert, S.; Riopel, D.; Abdul-Kader, W. A reverse logistics decisions conceptual framework. *Comput. Ind. Eng.* **2011**, *61*, 561–581. [CrossRef]
63. Kale, S.; Karaman Erkan, A. Evaluating the Knowledge Management Practices of Construction Firms by Using Importance–Comparative Performance Analysis Maps. *J. Constr. Eng. Manag.* **2011**, *137*, 1142–1152. [CrossRef]
64. Silva, R.V.; de Brito, J.; Dhir, R.K. Availability and processing of recycled aggregates within the construction and demolition supply chain: A review. *J. Clean. Prod.* **2017**, *143*, 598–614. [CrossRef]
65. Crowther, P. Design for Disassembly—Themes and Principles. *Environ. Des. Guide* **2005**, 1–7.

66. Foo, M.Y.; Kanapathy, K.; Zailani, S.; Shaharudin, M.R. Green purchasing capabilities, practices and institutional pressure. *Manag. Environ. Qual. Int. J.* **2019**, *30*, 1171–1189. [CrossRef]
67. Zhu, Q.; Sarkis, J.; Lai, K.-H. Institutional-based antecedents and performance outcomes of internal and external green supply chain management practices. *J. Purch. Supply Manag.* **2013**, *19*, 106–117. [CrossRef]
68. Banaeian, N.; Mobli, H.; Nielsen, I.E.; Omid, M. Criteria definition and approaches in green supplier selection—A case study for raw material and packaging of food industry. *Prod. Manuf. Res.* **2015**, *3*, 149–168. [CrossRef]
69. Shannak, R.O. Measuring knowledge management performance. *Eur. J. Sci. Res.* **2009**, *35*, 242–253.
70. El-Gohary, N.M.; Osman, H.; El-Diraby, T.E. Stakeholder management for public private partnerships. *Int. J. Proj. Manag.* **2006**, *24*, 595–604. [CrossRef]
71. Zhang, J.; Li, H.; Olanipekun, A.O.; Bai, L. A successful delivery process of green buildings: The project owners' view, motivation and commitment. *Renew. Energy* **2019**, *138*, 651–658. [CrossRef]
72. Olanipekun, A.O.; Chan, A.P.C.; Xia, B.; Ameyaw, E.E. Indicators of owner commitment for successful delivery of green building projects. *Ecol. Indic.* **2017**, *72*, 268–277. [CrossRef]
73. Tan, Y.t.; Shen, L.y.; Langston, C.; Liu, Y. Construction project selection using fuzzy TOPSIS approach. *J. Model. Manag.* **2010**, *5*, 302–315. [CrossRef]
74. Gharzeldeen, M.N.; Beheiry, S.M. Investigating the use of green design parameters in UAE construction projects. *Int. J. Sustain. Eng.* **2015**, *8*, 93–101. [CrossRef]
75. Hyde, R.; Watson, S.; Cheshire, W.; Thomson, M. *The Environmental Brief: Pathways for Green Design*; Taylor & Francis: London, UK, 2007.
76. Kozminska, U. Circular design: Reused materials and the future reuse of building elements in architecture. Process, challenges and case studies. *IOP Conf. Ser. Earth Environ. Sci.* **2019**, *225*, 012033. [CrossRef]
77. Khan, M.W.A.; Ting, N.H.; Kuang, L.C.; Darun, M.R.; Mehfooz, U.; Khamidi, M.F. Green Procurement in Construction Industry: A Theoretical Perspective of Enablers and Barriers. *MATEC Web Conf.* **2018**, *203*, 02012. [CrossRef]
78. Škapa, R.; Klapalová, A. Reverse logistics in Czech companies: Increasing interest in performance measurement. *Manag. Res. Rev.* **2012**, *35*, 676–692. [CrossRef]
79. Wong, J.K.W.; Chan, J.K.S.; Wadu, M.J. Facilitating effective green procurement in construction projects: An empirical study of the enablers. *J. Clean. Prod.* **2016**, *135*, 859–871. [CrossRef]
80. Shurrab, J.; Hussain, M.; Khan, M. Green and sustainable practices in the construction industry. *Eng. Constr. Archit. Manag.* **2019**, *26*, 1063–1086. [CrossRef]
81. Abraham, J.J.; Saravanakumar, R.; Ebenanjar, P.E.; Elango, K.S.; Vivek, D.; Anandaraj, S. An experimental study on concrete block using construction demolition waste and life cycle cost analysis. *Mater. Today Proc.* **2021**, in press. [CrossRef]

Article

Integrating BIM-Based LCA and Building Sustainability Assessment

José Pedro Carvalho [1,2,*], Ismael Alecrim [2], Luís Bragança [1,2,*] and Ricardo Mateus [1,2]

1 Institute for Sustainability and Innovation in Structural Engineering (ISISE), University of Minho, 4800-058 Guimarães, Portugal; ricardomateus@civil.uminho.pt
2 School of Engineering, University of Minho, 4800-058 Guimarães, Portugal; ismael_alecrim@hotmail.com
* Correspondence: jpcarvalho@civil.uminho.pt (J.P.C.); braganca@civil.uminho.pt (L.B.)

Received: 21 August 2020; Accepted: 7 September 2020; Published: 10 September 2020

Abstract: With the increasing concerns about building environmental impacts, building information modelling (BIM) has been used to perform different kinds of sustainability analysis. Among the most popular are the life cycle assessment (LCA) and building sustainability assessment (BSA). However, the integration of BIM-based LCA in BSA methods has not been adequately explored yet. This study addresses the relation between LCA and BSA within the BIM context for the Portuguese context. By performing an LCA for a Portuguese case study, a set of sustainability criteria from SBTool were simultaneous assessed during the process. The possibility of integrating BIM-based LCA into BSA methods can include more life cycle stages in the sustainability assessment and allow for normalising and producing more comparable results. BIM automates and connects different stages of the design process and provides information for multi-disciplinary data storage. However, there are still some constraints, such as different BSA/LCA databases and the necessity to manually introduce the embodied life cycle impacts of building materials. The scope of the BSA analysis can be expanded by integrating a complete LCA and be fostered by the support of BIM, effectively improving building sustainability according to local standards.

Keywords: building sustainability assessment (BSA); building information modelling (BIM); sustainability; life cycle assessment (LCA)

1. Introduction

The construction sector is highly accountable for several impacts on the environment [1,2]. Up to date, this sector is responsible for 40% of the EU energy demand, 36% of carbon emissions and 50% of raw material consumption [3]. With the relation between environmental impacts already been proved by the scientific community, authorities and general society are demanding more sustainable buildings [4].

Most of the building's life cycle impacts are a consequence of decisions made in the early design stages, making it extremely important to carefully select materials with low embodied impacts [5]. Researchers have already recognised the importance of early design stages to reduce buildings' life cycle environmental impacts and improve building sustainability [2,5,6]. Eleftheriadis et al. [7] have also identified that the early design phase is where benefits are more noticeable, as decisions cost less, are more effective and can be easier introduced. Thus, it is essential to act in such project stages to effectively reduce building environmental impacts.

Different methods and tools have been developed to evaluate buildings and other constructions' environmental performance. Among some of them, both building sustainability assessment (BSA) methods and life cycle assessment (LCA) tools have been extensively used [2,7–9]. The combination of such assessments can provide comprehensive data for designers to compare and select the best construction solutions and hence, developing more high-performance constructions. While BSA is intended to be a certification system which evaluates the building with a sustainable score, LCA usually focuses on the building elements' and materials' environmental impacts over the building life cycle [10]. Nevertheless, BSA methods often include a kind of LCA, either for the whole building or for its materials and components [11].

Due to the enormous pressure of the construction sector on the environment, LCA has been used to assess the overall building environmental impacts in recent years [3,7,11,12]. In 2010, Blengini and DiCarlo [13] had established that LCA was an appropriate method to assess the potential environmental impacts of the building sector. Their theory was shared and proved by different authors over the following years [14]. Despite the usefulness of LCA [15], there is still a need to consider the different aspects that can affect building performance. According to Vilches et al. [16], current research about LCA usually neglects social and cultural aspects, only focusing on the building energy consumption and carbon emissions. Nwodo and Anumba [17] concluded that to increase the usefulness of building LCA for decision-making, other multi-criteria assessment tools should also be included.

Furthermore, Hollberg et al. [18] suggest that LCA and sustainability certification should develop a common database for long-term use. As BSA methods assess multi-criteria features from a building and usually encompass a kind of LCA, the opportunity to combine LCA and BSA emerges. This relation will provide designers with a method to perform a broader and accurate analysis (considering social and economic aspects), reaching more significant overall results for society and the environment.

Due to the complexity in managing the vast quantity of data both to perform LCA and BSA [5,12,14,17,19], building information modelling (BIM) should be introduced to optimise designers' efforts and reduce process complexity [15–17,20,21]. The goal is also to improve the LCA performance and to collect enough data to perform both analyses during the early stages of a project, allowing for design guidance and optimisation [14]. The possibility to introduce different multi-disciplinary data into a single model makes BIM a useful platform for the comparison and introduction of sustainable measures in various project stages, especially in the early design phases [22].

Facing the existing opportunity, the aim of this study was to demonstrate the relationship between a BIM-based LCA and BSA for the Portuguese building context. By submitting a Portuguese dwelling case study to a BIM-based LCA process, the case study environmental impacts will be assessed, as well as a set of sustainability criteria from the BSA method SBTool. The research outcomes will establish a framework to carry out an LCA in combination with a BSA during the project's early stages, based on BIM methodology. Designers will be able to quickly assess their buildings' environmental impacts, while performing a concise sustainability assessment with few resources, addressing all the sustainability dimensions.

2. Literature Review

2.1. *Life Cycle Assessment (LCA)*

LCA is a commonly applied multi-disciplinary method to evaluate the environmental impacts of a product, process or activity [17,20,21]. Through the LCA process, the energy and material uses are identified and quantified through the whole product life cycle, including extraction, processing, manufacturing, transportation, use, reuse, maintenance, recycling or final disposal [5,17].

According to Nwodo and Anumba [12,15], the main objective of a building LCA concerns the minimisation of environmental impacts, carbon emissions, energy and costs. Besides the assessment of building environmental impacts in the project's early stages, LCA can also support decision making, by allowing the comparison of the embodied and the operational impact of different solutions [11,12,15,16]. LCA was already recognised as a critical tool to reduce buildings' environmental impacts and its use is continuously increasing [12,18,20,21]. As a result, in France and in the Netherlands, it is mandatory to apply a green building certification system, where LCA is often required [23].

LCA principles, framework, requirements and guidelines are defined in the ISO 14040:2006 and ISO 14044:2006 standards [15,24]. Under the construction scenario, LCA is oriented by the European Norms

15978 and 15804 [24], which have defined different regulations, analysis boundaries and modules according to the considered lifetime period. Up to date, the following modules are usually considered: Product/Manufacture stage (A1–A3), Construction process stage (A4–A5), Use (B1–B7), End-of-Life stage (C1–C4) and Benefits & Loads behind (D). The consideration of different modules/stages are defined in the boundaries of the analysis [24].

Different authors [7,15,16,24] have used ISO 14040 to encompass LCA framework into four distinct phases: goal and scope definition; life cycle inventory analysis, life cycle impact assessment and interpretation. In the first phase, the purpose of this study is defined, as well as the functional units and system boundaries. The second phase consists of gathering data related to the inputs/outputs of a product or process life cycle. Then, in the life cycle impact assessment, environmental impacts are quantified in different indicators, based on the inventory analysis. Finally, the last phase concerns the interpretation and analysis of impacts and the recommendations to improve the environmental performance.

Traditionally, buildings impacts are higher during the operational stage due to the significant energy demand of building integrated systems, lighting and appliances [1,16]. According to a review from Chau et al. [20], the operational building stage is the one that contributes the most to the building life cycle environmental impacts, followed by the structural materials. However, the relation between the embodied energy of materials and the operational energy is changing [1,11,16]. New buildings have less energy demand during the operational stage, and some recent studies showed that this stage accounts for about 60% of the whole life cycle impact [14,16]. Material-related impacts have increased their significance to 40%. Materials may be carefully faced in LCA, according to Häfliger et al. [25], as uncertainties related to building materials have important consequences on the final LCA result at the building scale.

Among the life cycle studies, two other approaches are mainly recognised by researchers [26]: life cycle energy assessment (LCEA) and life cycle carbon emissions assessment (LCCA). While the goal of LCEA is to reduce the primary energy use, by analysing the building energy inputs, the LCCA concerns the evaluation of carbon emissions as output over the building life cycle.

2.2. Building Information Modelling (BIM)

Facing the increasing complexity and size of construction projects, different technologies have been introduced to support designers in managing their projects and creating better buildings [2]. Among them, building information modelling (BIM) stands out as a working methodology, where all the project design and data are managed within a virtual model through the building life cycle [1,26].

BIM can improve process productivity, integrate multi-disciplinarily information into a single model and promote a collaborative environment throughout the project life cycle [3,22]. With stakeholders working in constant and real-time collaboration, errors, incompatibilities or omissions are usually avoided. Information exchange between stakeholders is generally made with industry foundation class (IFC) files, which contain building and construction industry data, and are normalised by the ISO 16739-1:2018 [27].

The application of the BIM method implies the development of a virtual object-oriented parametric model, which contains all the project data. According to the amount and type of data, the model level of development (LOD) is defined. The LOD specifies and articulates the content and reliability of a BIM model and ranges from 100—the conceptual model—to 500—the as-built model [28].

BIM can be used to enhance building sustainability and minimise errors through integrated design tools. According to Eleftheriadis et al. [7], the BIM contribution to sustainability assessment focuses on two perspectives: integrated project delivery and design optimisation. Moreover, they have concluded that the combination of BIM with sustainable strategies allows producing high-performance design alternatives. A similar conclusion was reached by Abanda and Byers [29] affirming that the possibility to simulate the building performance allows for the efficient development of high-performance buildings. Some of the most known applications of BIM for building sustainability are energy analysis, lightening and daylight analysis, estimation of water use, estimation of the renewable energy produced on-site, acoustic analysis, waste management, sustainability and life cycle assessment.

However, both the BIM method and the existing tools did not achieve their full potential for building sustainability yet [19,30]. Several authors argue that more sustainability issues should be considered in existing software and the interoperability between different software improved [31,32]. Stakeholders training and awareness for sustainability are also barriers to the broader implementation of BIM [33].

2.3. Building Sustainability Assessment (BSA)

For the past 20 years, different companies and organisations have been developing several building sustainability assessment (BSA) methods worldwide [30,34]. Despite the existence of several BSA methods adapted to each location, Leadership in Energy and Environmental Design (LEED), Building Research Establishment Environmental Assessment Method (BREEAM), and Sustainable Building Tool (SBTool) have been recognised as the basis for all the other approaches [35,36]. Nevertheless, Mahmoud et al. [10], argue that a common global method would be beneficial, as it would allow the comparisons between buildings form different locations. However, this approach would not consider the specific local aspects and conditions and non-consensus calculations would be required. Therefore, the researcher's tendency was to contextualise well known BSA methods to their specific regions of interest [37].

Overall, they intended to evaluate the specific buildings' features and aggregating all of them into a single sustainability score, according to the building location requirements [30]. They also encouraged the integration of sustainable measures, supported decision making and raised awareness of the building sector for sustainability issues [38,39].

To date, performing a BSA is considered a time-consuming and complex process, as multi-disciplinary data must be assessed and treated before and during the project phase [19]. Furthermore, it is based on an iterative process, and as project companies usually deal with strict deadlines, they often assess building sustainably in the latter stages, where modification costs are higher.

Facing the need to automate and integrate BSA during early project phases, the opportunity to take advantage of BIM capabilities arises. As a BIM model can store multi-disciplinary information and create specific sustainability properties, it allows to analyse and integrate different sustainability solutions with few resources [22].

From the three BSA methods mentioned above, SBTool is the only scheme that was adapted to the Portuguese scenario. Different adaptations were made for residential, office, healthcare buildings, schools, as well as for urban neighbourhoods [38,40,41]. In this study, the SBTool^{PT}-H version will be used, which is the Portuguese version for residential buildings. The aim of this method was to create a common methodology to assess the sustainability of Portuguese residential buildings and to demonstrate the benefits of adopting more sustainable solutions. In the SBTool^{PT}-H, there are 25 sustainability criteria sorted by three dimensions—environment, society and economy. The assessment procedure of each criterion is based on the comparison between the building performance and two benchmarks: the best and conventional national practices. After the assessment of all criteria, a weighting system is applied accordingly, and a sustainability score is obtained [38].

2.4. The Relation between BIM, LCA and BSA

The integration of the BSA and the LCA in the BIM process can significantly contribute to integrate sustainability assessment and LCA within the building sector [7]. Several studies have already been made on the integration of BIM in LCA and BIM in BSA. However, only a few have related the three approaches [42,43]. According to Carvalho et al. [8], BSA methods exploit the full potential of BIM, since it is necessary a set of multi-disciplinary criteria for their application. The same opinion is shared by Marrero et al. [34] for LCA, arguing that BIM allows to incorporate and extract those data from BIM.

BIM allows for relevant BSA credits to be directly calculated and documented [7]. Several authors have already used BIM to assess BSA criteria. A systematic review of Carvalho et al. [8] has analysed major publications addressing BIM and BSA, identifying LEED as the most assessed scheme and the energy and the material related as the most assessed categories. Azhar et al. [22] and Jalaei and Jrade [43,44] have focused their attention on LEED assessment with BIM-based procedures. Using different BIM software, Edwards et al. [30] have assessed eight credits from the BREEAM method, while Wong and Kuan [45] have gathered data for assessing 26 criteria from Building Environmental Assessment Method (BEAM) Plus. Gandhi and Jupp [46] have also applied BIM to assess 66% of the sustainability indicators of the Australian Green Star Building certification. Carvalho et al. [19] have proposed a methodologic BIM framework to assess 24 out of the 25 sustainability indicators of the Portuguese version of SBTool. All of them agreed that BIM allows for a faster sustainability assessment with fewer resources. As for the limitations, the authors pointed out the time-consuming and complex process, the need to use different software and interoperability gaps [19,22,31,33]. Moreover, they concluded the need to develop execution and coordination plans addressing building certification [39,45]. Chong et al. [31] have also proposed that future BIM standards should include requirements for a BSA.

BIM-based LCA is also an emerging trend [7]. Kreiner et al. [47] have created a BIM–LCA approach to improve building sustainability. Basbagill et al. [5] have developed a BIM framework to support the designer's decision making in the early project stages. By integrating BIM, LCA and other analysis, the impacts of different building designs were quickly compared. By assessing a Canadian residential building, Razaei et al. [1] have performed a full LCA. During the conceptual stage, a LOD 100 model was used, where uncertainties were given to materials. Then, in the design phase, the LCA was carried out with an LOD 300 model for more concise results. Rezaei et al. [1] agreed and stated that LCA should be applied at the conceptual design stage using an LOD 100, to introduce better decisions and decrease their environmental impacts. Sous-Verdaguer et al. [2] have also identified LOD 300 as the most appropriate to analyse environmental impacts during the early design stage. With a BIM-based method, Naneva et al. [23] have proposed a methodology to perform LCA in each building phase continuously. They have provided a decision-making support tool at the element and building level, where re-work is avoided. Despite all the applications, there are still some limitations on the relation between BIM and LCA, as interoperability issues, propensity for human error, license costs and the fact that the BIM model cannot store LCA data [3].

Typically, research on the integration of LCA in BIM focuses on extracting quantities to establish a Life Cycle Inventory. However, as usually, stakeholders do not have enough data to perform LCA in the early stages, only applying it once in the latter stages of a project [5,14,18]. To implement LCA in the project early stages, Rock et al. [14] have proposed a BIM-based LCA where designers can compare the embodied environmental impact of their solutions and effectively improve building design. A review study from Sous-Verdaguer et al. [2] identified three ways to link BIM and LCA: the quantification of materials and building elements (life cycle inventory—LCI); in addition to LCI, environmental information is integrated into BIM software, and; development of an automated process combining different data and software.

To date, it is easier to perform a BSA than a full LCA [11]. Although efforts were made to include LCA in BSA due to the need to simplify the implementation of an LCA [17], nowadays, certifications include LCA in their assessments as LEED, *Deutsche Gesellschaft für Nachhaltiges Bauen* (DGNB), *Haute Qualité Environnementale* (HQE), BREEAM or SBTool [11]. However, LCA in BSA is new, and there is a need to develop it further for better integration between LCA and global and local sustainability certification schemes [11]. Alshamrani et al. [48] integrated an LCA into LEED to improve sustainability assessment and support decision making for school buildings' structures and envelopes. A systematic review from Muller et al. [49] identified that BIM papers concerning building sustainability usually focus on the design stage, followed by the construction phase. The less addressed stage regards the final lifecycle phases. This leads to the comments by Elefteriadis et al. [7] highlighting the need to extend BIM use for sustainability purposes in order to maximise environmental performance and reach all the building life cycle stages. Therefore, the opportunity to explore the relation between BIM-based LCA and BSA emerges.

Jrade and Jalaei [43] have related a BIM-based LCA with the BSA method LEED. By generating and exporting quantities' take-offs from the BIM model with an external database (based on the Athena Impact Estimator tool), environmental impacts were re-evaluated, and LEED points were assessed. Roh et al. [42] performed a life cycle carbon emissions assessment and connected their results with the Korean Green Building Index (GBI).

3. Materials and Methods

This paper focuses on the relationship between building LCA and BSA based on the BIM method. By submitting a Portuguese dwelling case study to a BIM-based LCA process, the environmental impacts were assessed, as well as a set of sustainability criteria from the SBTool method.

To archive this goal, a Portuguese case study located in Porto (Portugal) was modelled and characterised in Autodesk Revit. Every building compartment was characterised with a room or space function, for the importing software to recognise the space's function and activity. The model was then exported to the Cype software environment via an IFC file through the BIMServer.center, which acts as an intermediary platform to use BIM models in the Cype environment. The model was used in Cypetherm REH to estimate the building energy consumption according to the Portuguese regulation. This was identified as the adequate software to calculate the energy performance of Portuguese buildings [19]. Cypetherm REH calculates the building primary energy consumption and its limit/reference value, according to Portuguese thermal regulation for residential buildings (REH). Primary energy calculation is based on conversion factors to convert final energy into primary energy. For instance, it indicates how much primary energy is used to generate a unit of electricity or a unit of useable thermal energy. According to the Portuguese regulation (Order No. 15793-D/2013), the conversion factors for Portugal were 2.5 kWh_{PE}/kWh for electricity and 1 kWh_{PE}/kWh for fuel.

After the energy analysis, the model was then analysed with the LCA software Tally, by using the available plugin for Autodesk Revit. After defining all the required data, such as the expected lifetime and the water/energy costs, the building life cycle environmental impacts were assessed. The building operational energy results (from Cypetherm REH) were included in the analysis.

The obtained results and quantities will be linked to the BSA method $SBTool^{PT}$-H, to automatically reach an assessment for criterion P1—construction materials embodied environmental impacts. However, during the LCA procedure, data to support the assessment of other sustainability criteria from $SBTool^{PT}$-H can also be collected. All the requirements will be identified to clearly establish and define the relation between a BIM-based LCA and the methodology of a sustainability assessment scheme.

The research procedure is summarised in Figure 1.

Regarding the case study, it was intended to be representative of Portuguese buildings but simple enough to perform a smooth analysis. A detached single-family dwelling of 90 m^2 was created, representing existing Portuguese buildings built at the end of the 20th century. Figure 2 presents the case study 3D model (from Autodesk Revit) and floor plan.

Construction solutions for the envelope and interior compartments (as well as their surface areas) were defined according to the conventional Portuguese practices and are described in Table 1. Insulation was added to meet the Portuguese thermal regulation.

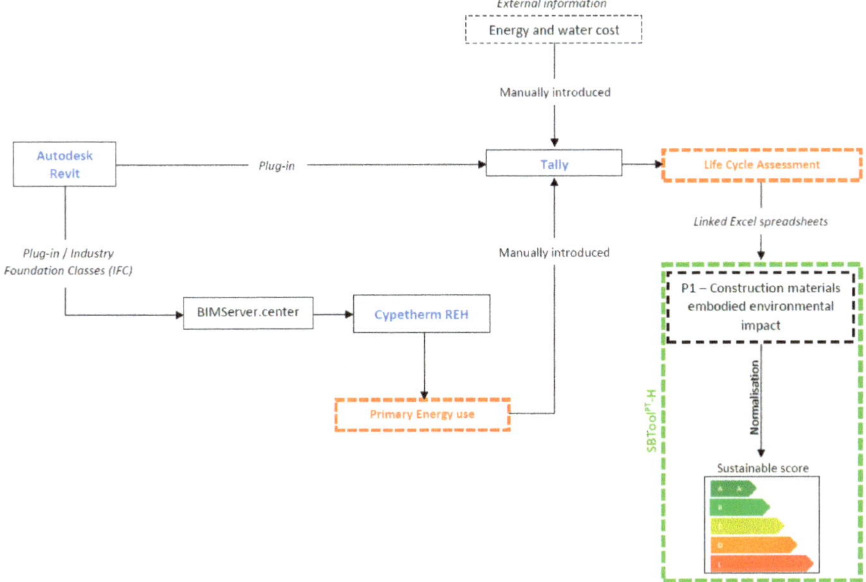

Figure 1. Research procedure.

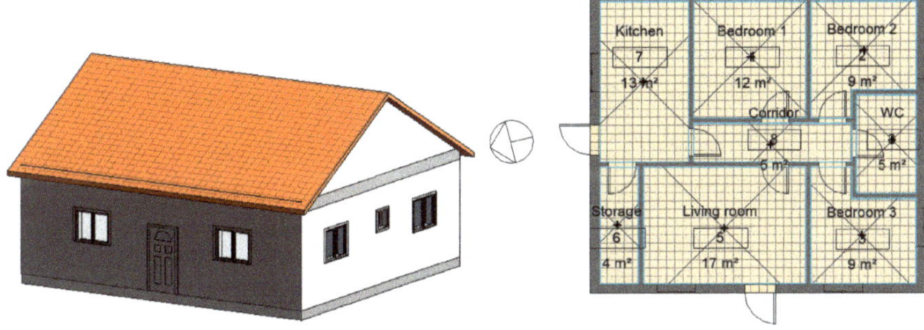

Figure 2. Case study model and floor plan.

Table 1. Case study construction characteristics.

Element	Surface Area (m^2)	Construction Solution
Exterior Walls	122.56	20 cm brick wall with external XPS insulation
Interior Walls	92.35	11 cm brick wall
Floor Slab	90.00	Concrete slab with internal XPS insulation and ceramic finishing
Roof Slab	90.05	Concrete slab with exterior XPS insulation
Roof	117.56	Ceramic Portuguese tile
Windows	7.56	Aluminium frame without thermal break and double glass
Doors	3.72	Exterior aluminium doors and interior wooden doors

4. Results

4.1. Cypetherm REH

The model was exported to the Cype environment via BIMServer.center, which has a specific plug-in (IFC export) for Autodesk Revit. The first step was to check and define the building envelope elements, interior elements, systems and project properties. Linear thermal bridges are automatically calculated by analysing the building elements' parametric relation.

By carrying out the energy performance simulation, the Primary Energy (PE) use of the building was reached, including winter, summer and Domestic Hot Water (DHW) needs. According to the Portuguese regulation, the annual required primary energy demand for the case study is 9840.17 kWh. Table 2 presents a summary of the results. For the annual energy simulation, the remaining aspects were considered:

- Building occupancy of four people;
- Solar collector able to produce 1280 kWh/year for DHW with a natural gas backup system;
- Air renovations (0.6 per hour for summer and 0.4 for winter).

Table 2. Energy performance simulation results.

Heating Needs (kWh/year)	Cooling Needs (kWh/year)	Domestic Hot Water (DHW) Needs (kWh/year)	Primary Energy Needs (kWh_{PE}/year)	Regulation's Limit for the Primary Energy Needs (kWh_{PE}/year)
Electricity 3388.26	Electricity 323.22	Natural gas 920.11	9840.17	16,696.11

These data will be further used to carry out the LCA in Tally. However, with this energy assessment procedure, designers can already gather data to assess three other criteria from SBTool^{PT}-H, namely:

- Energy efficiency category
    - P7—Primary energy need
        - Required data:
            - Building primary energy needs (and regulation limit);
            - Building compartments/total area.
    - P8—On-site energy production from renewables
        - Required data:
            - Building primary energy needs (and regulation limit), cooling, heating and DHW needs;
            - Renewable energy production;
            - Building compartments/total area;
            - Number of occupants.
- Occupant's health and comfort
    - P19—Natural light performance
        - Required data:
            - Visible sky angle, given by the horizon and horizontal obstruction angle (which are automatically calculated by Cypetherm REH according to the building geometry);

- Building interior surface area (including glazed area).

Both the information from the BIM model and the simulation in Cypetherm REH can provide all these data with an exception for renewable energy production. For this research, a spreadsheet for the renewable energy estimation provided by the Portuguese Directorate-General for Energy and Geology (DGEG) was used.

4.2. *Tally*

With the building energy demand, the LCA simulation was carried out in Tally. The existing plug-in for Autodesk Revit was used for the analysis.

After selecting the type of analysis (full building assessment), the included categories and life cycle stages, materials were linked with the Tally material database—GaBi LCI databases—to gather their associated impacts. The building operational energy was introduced using the Cypetherm REH results—3711 kWh from electricity (heating and cooling needs) and 920 kWh from natural gas (DHW needs).

A 60 year lifetime was considered for the analysis. The boundaries were defined to include all life cycle stages (cradle-to-grave), including material manufacturing, maintenance, replacement and end of life.

The results from Tally are expressed in environmental impact categories, which translates all emissions and fuel use into quantities of categorised environmental impacts. The following impacts were considered: acidification potential (AP), eutrophication potential (EP), global warming potential (GWP), ozone depletion potential (ODP), smog formation potential (SFP), primary energy demand (PED), non-renewable energy demand (NRED) and renewable energy demand (RED).

The achieved impacts per life cycle stage for the case study are presented in Table 3.

Table 3. Environmental impacts per life cycle stage.

Environmental Impact Totals	Product Stage (A1–A3)	Construction Stage (A4)	Use Stage (B2–B6)	End of Life Stage (C2–C4)	Module D (D)
Global Warming (kg CO_2eq)	5.18×10^4	1.62×10^3	1.06×10^5	4.27×10^3	-1.86×10^3
Acidification (kg SO_2eq)	1.30×10^2	7.52	2.13×10^2	19.8	−6.95
Eutrophication (kg Neq)	8.85	6.12×10^{-1}	22.5	1.52	-1.46×10^{-1}
Smog Formation (kg O_3eq)	2.24×10^3	2.49×10^2	3.49×10^3	3.62×10^2	−36.4
Ozone Depletion (kg CFC-11eq)	3.67×10^{-5}	5.56×10^{-11}	5.44×10^{-5}	7.13×10^{-10}	8.17×10^{-6}
Primary Energy (MJ)	7.48×10^5	2.36×10^4	2.54×10^6	6.64×10^4	-2.31×10^4
Non-Renewable Energy (MJ)	6.91×10^5	2.30×10^4	1.42×10^6	6.21×10^4	-1.86×10^4
Renewable Energy (MJ)	5.76×10^4	5.71×10^2	1.12×10^6	4.38×10^3	-4.50×10^3

Figure 3 shows the same impacts in percentages to fully understand the building's major impacts. The operational building stage (B6) is the major contributor to the building impacts, followed by the product stage (A1–A4). The end of life (C2–C4) and module D (D) stages are the less significant ones.

Tally can also provide the results per material, which are presented in Figure 4. As it is possible to understand, both the building structure (concrete) and walls (masonry) are the main contributors to environmental impacts. The building openings and glazing, as well as building finishes, are the materials which contribute less for the building's environmental impacts.

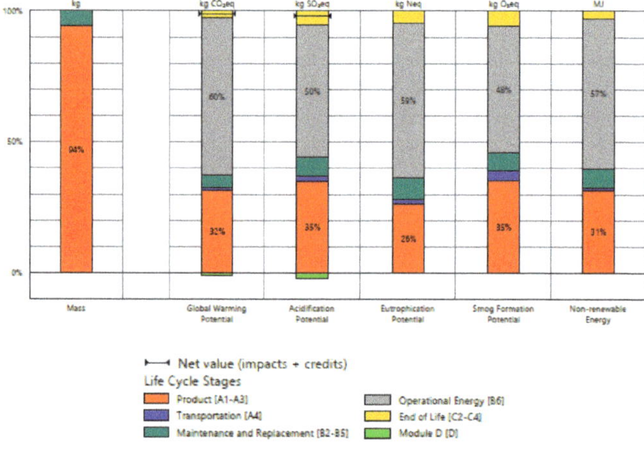

Figure 3. Environmental impacts per life cycle stage.

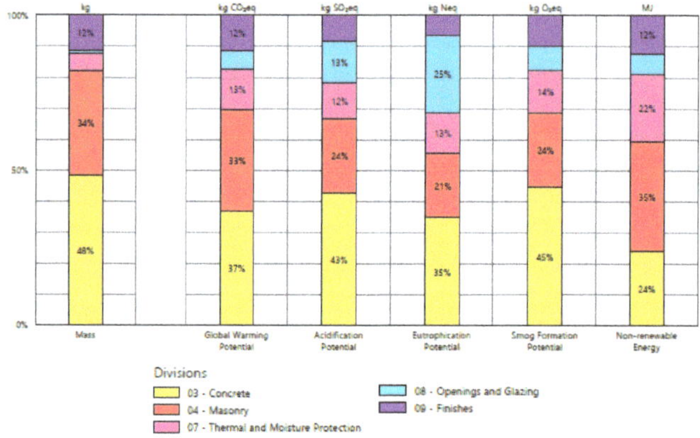

Figure 4. Environmental impacts per material type—Tally.

4.3. Sustainability Assessment

Tally results were linked to the SBTool^{PT}-H spreadsheet to reach an assessment for criterion P1-Construction materials embodied environmental impacts. The Tally list of quantities was used together with the SBTool materials environmental impacts database to reach a faster assessment. As a first step, the assessment procedure requires the calculation of environmental impacts by multiplying the quantities of the materials with the SBTool^{PT}-H database impact factors (Table 4). The following elements were considered for the analysis: Exterior and interior walls, envelope openings, floor slab and roof. Unlike Tally, SBTool^{PT}-H does not consider interior openings.

Then, benchmarks for the best conventional practice are calculated, based on the building elements area (Table 5):

- Conventional practice benchmark—impact factors pre-defined in SBTool^{PT}-H adapted for the Portuguese region. It is given by multiplying the element areas by those factors;
- Best practice benchmark—25% of the conventional practice.

At the end, the building performance will be faced with both benchmarks, and a normalised score for each environmental impact category is computed. By applying a weighting system, which was defined according to the Portuguese standards and environmental, societal and economic contexts, the final quantitative score for P1 is assessed. The normalisation procedure is presented in Table 6.

Table 4. Environmental impacts per life cycle stage.

Solution			Area (m^2)	Environmental Impact Categories Quantification (per m^2)						Environmental Impact Categories Quantification					
				GWP	ODP	AP	POCP	EP	FFDP	GWP	ODP	AP	POCP	EP	FFDP
				($kgCO_2$)	(kgCFC-11)	($kgSO_2$)	(kgC_2H_4)	($kgPO_4$)	(MJ)	($kgCO_2$)	(kgCFC-11)	($kgSO_2$)	(kgC_2H_4)	($kgPO_4$)	(MJ)
Exterior Walls		Common brick 20 cm	122.56	2.20×10^{-1}	1.58×10^{-8}	5.48×10^{-4}	4.00×10^{-5}	6.71×10^{-5}	2.58	27.00	1.94×10^{-6}	6.72×10^{-2}	4.90×10^{-3}	8.22×10^{-3}	3.16×10^{-2}
		XPS 5 cm	122.56	4.14	1.10×10^{-7}	1.49×10^{-2}	6.75×10^{-3}	1.24×10^{-3}	1.05×10^{2}	5.07×10^{2}	1.35×10^{-5}	1.83	8.27×10^{-1}	1.52×10^{-1}	1.29×10^{4}
		Cement plaster in both sides (2 cm each)	122.56	1.95×10^{-1}	8.00×10^{-9}	3.15×10^{-4}	1.29×10^{-5}	4.87×10^{-5}	1.31	23.90	9.80×10^{-7}	3.86×10^{-2}	1.58×10^{-3}	5.97×10^{-3}	1.61×10^{2}
Interior Walls		Common brick 11 cm	92.35	2.20×10^{-1}	1.58×10^{-8}	5.48×10^{-4}	4.00×10^{-5}	6.71×10^{-5}	2.58	20.30	1.46×10^{-6}	5.06×10^{-2}	3.69×10^{-3}	6.20×10^{-3}	2.38×10^{2}
		Cement plaster in both sides (2 cm each)	92.35	1.95×10^{-1}	8.00×10^{-9}	3.15×10^{-4}	1.29×10^{-5}	4.87×10^{-5}	1.31	18.00	7.39×10^{-7}	2.91×10^{-2}	1.19×10^{-3}	4.50×10^{-3}	1.21×10^{2}
Floor Slab	Finishing	Mosaic floor	90.00	7.63×10^{-1}	8.16×10^{-8}	2.93×10^{-3}	1.36×10^{-4}	2.75×10^{-4}	14.00	68.70	7.34×10^{-6}	2.64×10^{-1}	1.22×10^{-2}	2.48×10^{-2}	1.26×10^{3}
		Plaster 2 cm	90.00	1.95×10^{-1}	8.00×10^{-9}	3.15×10^{-4}	1.29×10^{-5}	4.87×10^{-5}	1.31	17.60	7.20×10^{-7}	2.84×10^{-2}	1.16×10^{-3}	4.38×10^{-3}	1.18×10^{2}
		XPS 5 cm	90.00	4.14	1.10×10^{-7}	1.49×10^{-2}	6.75×10^{-3}	1.24×10^{-3}	1.05×10^{2}	3.73×10^{2}	9.90×10^{-6}	1.34	6.08×10^{-1}	1.12×10^{-1}	9.45×10^{3}
		Reinforced concrete 20 cm	90.00	1.48×10^{-1}	3.55×10^{-9}	5.56×10^{-4}	5.28×10^{-5}	5.76×10^{-5}	1.24	13.30	3.20×10^{-7}	5.00×10^{-2}	4.75×10^{-3}	5.18×10^{-3}	1.12×10^{2}
Openings		Glass	11.43	9.73×10^{-1}	8.01×10^{-8}	8.51×10^{-3}	2.86×10^{-4}	6.53×10^{-4}	11.50	11.10	9.16×10^{-7}	9.73×10^{-2}	3.27×10^{-3}	7.46×10^{-3}	1.31×10^{2}
		Aluminium	51.20	4.28	1.84×10^{-6}	3.80×10^{-2}	2.23×10^{-3}	1.21×10^{-3}	68.20	2.19×10^{2}	9.42×10^{-5}	1.95	1.14×10^{-1}	6.20×10^{-2}	3.49×10^{3}
Roof	Finishing	Ceramic Portuguese tile	117.56	8.16×10^{-1}	8.41×10^{-8}	2.90×10^{-3}	1.55×10^{-4}	2.85×10^{-4}	14.60	95.90	9.89×10^{-6}	3.41×10^{-1}	1.82×10^{-2}	3.35×10^{-2}	1.72×10^{3}
		XPS 8 cm	90.05	4.14	1.10×10^{-7}	1.49×10^{-2}	6.75×10^{-3}	1.24×10^{-3}	1.05×10^{2}	3.73×10^{2}	9.90×10^{-6}	1.34	6.08×10^{-1}	1.12×10^{-1}	9.45×10^{3}
		Flexible membrane with bitumen	90.05	5.81×10^{-1}	7.27×10^{-7}	7.27×10^{-3}	1.94×10^{-3}	3.02×10^{-4}	5.33×10^{1}	5.23×10^{1}	6.55×10^{-5}	6.55×10^{-1}	1.75×10^{-1}	2.72×10^{-2}	4.80×10^{3}
		Steam polyvinyl chloride (PVC) barrier	90.05	1.97	2.84×10^{-9}	5.35×10^{-3}	3.12×10^{-4}	7.59×10^{-4}	46.90	1.77×10^{2}	2.56×10^{-7}	4.82×10^{-1}	2.81×10^{-2}	6.83×10^{-2}	4.22×10^{3}
		Light concrete (5 cm)	90.05	1.10×10^{-1}	3.55×10^{-9}	1.79×10^{-4}	6.49×10^{-6}	2.84×10^{-5}	5.56×10^{-1}	9.90	3.20×10^{-7}	1.61×10^{-2}	5.84×10^{-4}	2.56×10^{-3}	50.10
		Reinforced concrete 20 cm	90.05	1.48×10^{-1}	3.55×10^{-9}	5.56×10^{-4}	5.28×10^{-5}	5.76×10^{-5}	1.24	13.30	3.20×10^{-7}	5.01×10^{-2}	4.75×10^{-3}	5.19×10^{-3}	1.12×10^{2}
		Plaster 2 cm	90.05	1.95×10^{-1}	8.00×10^{-9}	3.15×10^{-4}	1.29×10^{-5}	4.87×10^{-5}	1.31	17.60	7.20×10^{-7}	2.84×10^{-2}	1.16×10^{-3}	4.39×10^{-3}	1.18×10^{2}
Total life cycle environmental impacts										2.04×10^{3}	2.19×10^{-4}	8.65	2.42	6.45×10^{-1}	4.87×10^{4}

Table 5. Benchmarks for the analysis.

Solution Type	Area (m^2)	Environmental Impact Categories Quantification (per m^2)						Environmental Impact Categories Quantification					
		GWP	ODP	AP	POCP	EP	FFDP	GWP	ODP	AP	POCP	EP	FFDP
		($kgCO_2$)	(kgCFC-11)	($kgSO_2$)	(kgC_2H_4)	($kgPO_4$)	(MJ)	($kgCO_2$)	(kgCFC-11)	($kgSO_2$)	(kgC_2H_4)	($kgPO_4$)	(MJ)
Exterior Walls	122.56	56.40	3.54×10^{-6}	1.52×10^{-1}	1.64×10^{-2}	1.95×10^{-2}	5.84×10^2	6.91×10^3	4.34×10^{-4}	18.60	2.01	2.39	7.16×10^4
Interior Walls	92.35	28.90	1.80×10^{-6}	6.52×10^{-2}	3.77×10^{-3}	9.24×10^{-3}	2.34×10^2	2.67×10^3	1.66×10^{-4}	6.02	3.48×10^{-1}	8.53×10^{-1}	2.16×10^4
Floor Slab	90.00	70.50	3.51×10^{-6}	1.73×10^{-1}	6.21×10^{-3}	2.75×10^{-2}	4.65×10^2	6.35×10^3	3.16×10^{-4}	15.60	5.59×10^{-1}	2.48	4.19×10^4
Floor Slab—Finishes	90.00	9.73	8.19×10^{-7}	2.97×10^{-2}	1.32×10^{-3}	3.30×10^{-3}	1.12×10^2	8.76×10^2	7.37×10^{-5}	2.67	1.19×10^{-1}	2.97×10^{-1}	1.01×10^4
Envelope Openings	11.28	8.31	1.17×10^{-6}	1.16×10^{-1}	2.29×10^{-3}	8.18×10^{-3}	1.04×10^3	93.70	1.32×10^{-5}	1.31	2.58×10^{-2}	9.23×10^{-2}	1.17×10^4
Roof	90.05	71.30	3.60×10^{-6}	1.43×10^{-1}	6.73×10^{-3}	2.46×10^{-2}	4.51×10^2	6.42×10^3	3.24×10^{-4}	12.90	6.06×10^{-1}	2.22	4.06×10^4
Roof—Finishes	117.56	16.70	1.15×10^{-6}	4.06×10^{-2}	2.93×10^{-3}	4.82×10^{-3}	1.64×10^2	1.96×10^3	1.35×10^{-4}	4.77	3.44×10^{-1}	5.67×10^{-1}	1.93×10^4
Total life-cycle environmental—Conventional practice								2.53×10^4	1.46×10^{-3}	61.90	4.01	8.89	2.17×10^5
Total life-cycle environmental—Best practice								6.32×10^3	3.66×10^{-4}	15.50	1.00	2.22	5.42×10^4

Table 6. Environmental impacts normalisation.

Environmental Impact Categories	Life Cycle Impacts (per m^2 and per Year)				Environmental Impact Category Weight (%) (B)	Weighted Value = (A) × (B)
	Best Practice	Conventional Practice	Case Study Performance	Normalised Value (A)		
GWP ($KgCO_2$)	6.32×10^3	2.53×10^4	2.04×10^3	1.23	40.7	0.499
ODP (kgCFC-11)	3.66×10^{-4}	1.46×10^{-3}	2.19×10^{-4}	1.13	8.4	0.095
AP ($KgSO_2$)	15.50	61.90	8.65	1.15	13.6	0.156
POCP ($kg.C_2H_4$)	1.00	4.01	2.42	0.53	10.1	0.054
EP (kg PO_4)	2.22	8.89	6.45×10^{-1}	1.24	13.6	0.168
FFDP (MJ)	5.42×10^4	2.17×10^5	4.87×10^4	1.03	13.6	0.141
			$\Sigma \cdot =$ ·Environmental performance $\cdot \left(\overline{P_{LCA}}\right)$			1.112

According to the SBTool^{PT}-H assessment scheme, the case study has reached a score of A+ in criterion P1—construction materials embodied environmental impacts, which is above national best practices (Table 7 converts the quantitative score into a qualitative score).

Table 7. SBTool^{PT}-H P1 final score.

Qualitative Level	Quantitative Value	Score
A+	$\overline{P_{LCA}} > 1.00$	X
A	$0.70 < \overline{P_{LCA}} \leq 1.00$	
B	$0.40 < \overline{P_{LCA}} \leq 0.70$	
C	$0.10 < \overline{P_{LCA}} \leq 0.40$	
D	$0.00 < \overline{P_{LCA}} \leq 0.10$	
E	$\overline{P_{LCA}} < 0.00$	

5. Discussion

The applied procedure has related BIM-based LCA with the assessment of building sustainability schemes. As demonstrated, to perform an LCA for the Portuguese context, building operational energy must be previously estimated according to the Portuguese standards. The Cypetherm REH was used to conduct a concise energy performance simulation accordingly to the Portuguese thermal regulation—REH. Results have shown a common trend in Portugal, with higher heating demand. Despite the usefulness of Cypetherm REH for energy performance simulation in Portugal, it is not able to estimate renewable energy production, a mandatory parameter for the energy performance characterisation. The estimation of the on-site renewable energy production was made externally and introduced in Cypetherm REH. Besides the calculation of the building operational energy demand (for the LCA), the use of Cypetherm REH also provided the required data to assess a set of other sustainability criteria from SBTool^{PT}-H. Before the simulation itself, the software automatically determines the obstruction and horizon angles for windows, based on the parametric building geometry and surroundings (both made in Autodesk Revit). This information, together with the building and glazed area, can be used to fully assess criterion P19—natural light performance. After the energy simulation, results can be used to fully assess energy efficiency category criteria P7—primary energy need, and P8—on-site energy production from renewables. Overall, Cypetherm REH can provide results to assess three SBTool^{PT}-H criteria and data to support the LCA. Cype environment also allows creating BIM models. However, Autodesk Revit was selected as it is commonly used by researchers [8] and it encompasses a plug-in to export IFC files for the Cype environment.

Regarding the LCA, Tally plug-in for Autodesk Revit was used to export the building geometry and quantities to Tally. This software recognises the building elements according to the building parametric relation and materials/elements classes. To carry out the simulation, BIM model materials are linked with a Tally database (GaBi) to reach their environmental impacts. Achieved results meet other research conclusions by pointing out the building use stage as the most critical one, followed by the product stage. The same conclusions were made for the materials impacts, highlighting the negative contribution of concrete elements (building structure) and masonry units (for all the environmental impacts and mass). According to the analysis, the building's finishing materials are the most environmentally friendly, with fewer environmental impacts, while the building openings and glazed area have the lowest mass. The Tally analysis provides a full environmental impacts report, as well as a material inventory spreadsheet which can be used to export and link building material quantities. Generally, Tally allows for a faster and intuitive analysis, but the need to associate building materials with its database hinders the assessment procedure. Note that Tally is adapted to the United States region, and the material impacts are related to US's common practices [2]. Nevertheless, the procedure to relate BIM-based LCA with BSA remains the same.

The Tally material spreadsheet allows for a direct assessment of the SBTool^{PT}-H criterion P1 by proving the required quantities for the evaluation. However, building materials still must be matched with the Tally database, slowing the assessment process. Despite the good result achieved in the sustainability assessment, results should be carefully interpreted. SBTool analysis only focuses on the product stage, and some building features were not considered. Water and wastewater infrastructures were not modelled and building interior and exterior painting were not considered in both analyses (Tally and SBTool). When added, these features will significantly increase the building impact, resulting in a less positive sustainability grade. The comparison between SBTool and Tally results do not have a common path for comparisons. Besides the focus of SBTool only on the product stage, environmental impact databases are different. If similar databases were used, the assessment process could be improved and provide more comparable results. However, these databases must be region-oriented, according to the BSA scope.

Overall, the process to carry out a BIM-based LCA for the Portuguese context requires the use of different software and data, which can support the assessment of BSA. During the LCA procedure, data to fully support the evaluation of 4 SBTool^{PT}-H criteria can be quickly gathered. By relating LCA and BSA, building sustainability can be easily and faster evaluated with more complete and realistic results.

As the LCA directly interferes with BSA, its inclusion in the sustainability evaluation should take part in the assessment process. Thus, BSA criteria related to LCA can evaluate more life cycle stages with more complete and detailed data, promoting BSA methods' reliability. The use of BSA methods also facilitates, normalises and levels LCA results, for a more straightforward interpretation and comparison between buildings through BSA results. The use of BIM automates the whole process and allows for proper the management of input and output data. It also provides for a faster evaluation due to its interoperability capabilities and for multi-disciplinary data storage, which is essential to perform a different kind of sustainability analysis.

6. Conclusions

With the increasing demand for more sustainable buildings, new methods and approaches to design and build must be developed. The emergence of BIM in the construction industry has raised the awareness of researchers to optimise design procedures, allowing for time and resources saving while producing high-performance buildings. The application of sustainability tools, such as life cycle assessment and building sustainability assessment, has also gained new momentum and attractiveness in the scope of BIM. The interaction between LCA, BSA and BIM can be extremely valuable for a proper interpretation of data, to provide a complete sustainability analysis and to avoid re-work.

This research has demonstrated the relation between BIM–LCA and building sustainability assessment for a Portuguese case study. It allowed for the development of the current knowledge on LCA and BSA integration, as well as to gather more specific oriented and complete data to improve building sustainability. Moreover, it proved that LCA should be integrated with BSA analysis, as it directly provides data to assess a set of sustainability criteria. For the SBTool^{PT}-H case, the LCA also provides a cradle-to-grave analysis, which can widen the actual boundary that is focused on the product stage. When designing a sustainable building, this relation can significantly save designers time and support their decisions with more comparable results. The multi-disciplinary data storage of BIM and its interoperability capabilities also allow gathering data for other sustainability analysis.

This study has also identified some existing constraints which must be approached. The main barrier concerns databases, which are different among BSA methods and LCA tools. If identical databases were used, the evaluation process can run smoother and provide more direct and comparable results. Additionally, some databases would also allow for automatic material recognition to assign and calculate the potential environmental impacts. This will be an important improvement since, at the moment, it is a manual process that is necessary to be conducted by the sustainability evaluator.

The integration of LCA and BSA and its assessment with BIM can enhance the usefulness and scope of these sustainability tools. It creates the opportunity to optimise the evaluation procedure, to make decisions with more support data and to simplify the interpretation of results. Together they can effectively improve the sustainability of the built environment considering local standards and trends.

Author Contributions: Conceptualization, J.P.C. and I.A.; methodology, J.P.C. and I.A.; validation, L.B. and R.M.; formal analysis, J.P.C. and I.A.; investigation, J.P.C.; writing—original draft preparation, J.P.C.; writing—review and editing, L.B. and R.M.; supervision, L.B. and R.M.; funding acquisition, J.P.C. All authors have read and agreed to the published version of the manuscript.

Funding: This research was funded by the Portuguese Foundation for Science and Technology, grant number SFRH/BD/145735/2019.

Conflicts of Interest: The authors declare no conflict of interest.

References

1. Rezaei, F.; Bulle, C.; Lesage, P. Integrating Building Information Modeling and Life Cycle Assessment in the Early and Detailed Building Design Stages. *Build. Environ.* **2019**, *153*, 158–167. [CrossRef]
2. Santos, R.; Costa, A.A.; Silvestre, J.D.; Pyl, L. Integration of LCA and LCC Analysis Within a BIM-Based Environment. *Autom. Constr.* **2019**, *103*, 127–149. [CrossRef]
3. Soust-Verdaguer, B.; Llatas, C.; García-Martínez, A. Critical review of bim-based LCA method to buildings. *Energy Build.* **2017**, *136*, 110–120. [CrossRef]
4. Li, Y.; Chen, X.; Wang, X.; Xu, Y.; Chen, P.-H. A Review of Studies on Green Building Assessment Methods by Comparative Analysis. *Energy Build.* **2017**, *146*, 152–159. [CrossRef]
5. Basbagill, J.; Flager, F.L.; Lepech, M.; Fischer, M. Application of Life-Cycle Assessment to Early Stage Building Design for Reduced Embodied Environmental Impacts. *Build. Environ.* **2013**, *60*, 81–92. [CrossRef]
6. Azhar, S.; Brown, J.W.; Sattineni, A. A Case Study of Building Performance Analyses Using Building Information Modeling. In Proceedings of the 27th International Symposium on Automation and Robotics in Construction (ISARC-27), Bratislava, Slovakia, 25–27 June 2010; pp. 213–222.
7. Eleftheriadis, S.; Mumovic, D.; Greening, P. Life Cycle Energy Efficiency in Building Structures: A Review of Current Developments and Future Outlooks Based on BIM Capabilities. *Renew. Sustain. Energy Rev.* **2017**, *67*, 811–825. [CrossRef]
8. Carvalho, J.P.; Braganca, L.; Mateus, R. A Systematic Review of the Role of BIM in Building Sustainability Assessment Methods. *Appl. Sci.* **2020**, *10*, 4444. [CrossRef]
9. Ansah, M.K.; Chen, X.; Yang, H.; Lu, L.; Lam, P.T. A Review and Outlook for Integrated BIM Application in Green Building Assessment. *Sustain. Cities Soc.* **2019**, *48*, 101576. [CrossRef]
10. Mahmoud, S.; Zayed, T.; Fahmy, M. Development of Sustainability Assessment Tool for Existing Buildings. *Sustain. Cities Soc.* **2019**, *44*, 99–119. [CrossRef]
11. Anand, C.K.; Amor, B. Recent Developments, Future Challenges and New Research Directions in LCA of Buildings: A Critical Review. *Renew. Sustain. Energy Rev.* **2017**, *67*, 408–416. [CrossRef]
12. Rashid, A.F.A.; Yusoff, S. A Review of Life Cycle Assessment Method for Building Industry. *Renew. Sustain. Energy Rev.* **2015**, *45*, 244–248. [CrossRef]
13. Blengini, G.A.; Di Carlo, T. Energy-Saving Policies and Low-Energy Residential Buildings: An LCA Case Study to Support Decision Makers in Piedmont (Italy). *Int. J. Life Cycle Assess.* **2010**, *15*, 652–665. [CrossRef]
14. Röck, M.; Hollberg, A.; Habert, G.; Passer, A. LCA and BIM: Visualization of Environmental Potentials in Building Construction at Early Design Stages. *Build. Environ.* **2018**, *140*, 153–161. [CrossRef]
15. Soust-Verdaguer, B.; Llatas, C.; Garcia-Martinez, A. Simplification in Life Cycle Assessment of Single-Family Houses: A Review of Recent Developments. *Build. Environ.* **2016**, *103*, 215–227. [CrossRef]
16. Vilches, A.; Garcia-Martinez, A.; Sanchez-Montañes, B. Life Cycle Assessment (LCA) of Building Refurbishment: A Literature Review. *Energy Build.* **2017**, *135*, 286–301. [CrossRef]
17. Nwodo, M.N.; Anumba, C.J. A Review of Life Cycle Assessment of Buildings Using a Systematic Approach. *Build. Environ.* **2019**, *162*, 106290. [CrossRef]
18. Hollberg, A.; Genova, G.; Habert, G. Evaluation of BIM-Based LCA Results for Building Design. *Autom. Constr.* **2020**, *109*, 102972. [CrossRef]
19. Carvalho, J.P.; Braganca, L.; Mateus, R. Optimising Building Sustainability Assessment Using BIM. *Autom. Constr.* **2019**, *102*, 170–182. [CrossRef]

20. Chau, C.; Leung, T.; Ng, W. A Review on Life Cycle Assessment, Life Cycle Energy Assessment and Life Cycle Carbon Emissions Assessment on Buildings. *Appl. Energy* **2015**, *143*, 395–413. [CrossRef]
21. Sharma, A.; Saxena, A.; Sethi, M.; Shree, V.; Goel, V. Life Cycle Assessment of Buildings: A Review. *Renew. Sustain. Energy Rev.* **2011**, *15*, 871–875. [CrossRef]
22. Azhar, S.; Carlton, W.A.; Olsen, D.; Ahmad, I. Building Information Modeling for Sustainable Design and LEED® Rating Analysis. *Autom. Constr.* **2011**, *20*, 217–224. [CrossRef]
23. Naneva, A.; Bonanomi, M.; Hollberg, A.; Habert, G.; Hall, D. Integrated BIM-Based LCA for the Entire Building Process Using an Existing Structure for Cost Estimation in the Swiss Context. *Sustainability* **2020**, *12*, 3748. [CrossRef]
24. Köseci, F.C. Integrated Life Cycle Assessment (LCA) to Building Information Modelling (BIM). Master's Thesis, KTH Royal Institute of Technology, Stockholm, Sweden, 2018. Available online: https://www.diva-portal.org/smash/get/diva2:1229841/FULLTEXT01.pdf (accessed on 21 July 2020).
25. Häfliger, I.-F.; John, V.; Passer, A.; Lasvaux, S.; Hoxha, E.; Saade, M.R.M.; Habert, G. Buildings Environmental impacts' Sensitivity Related to LCA Modelling Choices of Construction Materials. *J. Clean. Prod.* **2017**, *156*, 805–816. [CrossRef]
26. Succar, B. Building Information Modelling Framework: A Research and Delivery Foundation for Industry Stakeholders. *Autom. Constr.* **2009**, *18*, 357–375. [CrossRef]
27. El-Diraby, T.E.; Krijnen, T.; Papagelis, M. BIM-Based Collaborative Design and Socio-Technical Analytics of Green Buildings. *Autom. Constr.* **2017**, *82*, 59–74. [CrossRef]
28. AIA. AIA Document G202TM-2013: Project Building Information Modeling Protocol Form. Available online: http://architectis.it/onewebmedia/AIA%C2%AE%20Document%20G202TM%20%E2%80%93%202013.pdf (accessed on 7 September 2020).
29. Abanda, F.; Byers, L. An Investigation of the Impact of Building Orientation on Energy Consumption in a Domestic Building Using Emerging BIM (Building Information Modelling). *Energy* **2016**, *97*, 517–527. [CrossRef]
30. Edwards, R.E.; Lou, E.C.; Bataw, A.; Kamaruzzaman, S.N.; Johnson, C. Sustainability-Led Design: Feasibility of Incorporating Whole-Life Cycle Energy Assessment into BIM for Refurbishment Projects. *J. Build. Eng.* **2019**, *24*, 100697. [CrossRef]
31. Chong, H.-Y.; Lee, C.-Y.; Wang, X. A Mixed Review of the Adoption of Building Information Modelling (BIM) for Sustainability. *J. Clean. Prod.* **2017**, *142*, 4114–4126. [CrossRef]
32. Liu, Z.; Osmani, M.; Demian, P.; Baldwin, A. A BIM-Aided Construction Waste Minimisation Framework. *Autom. Constr.* **2015**, *59*, 1–23. [CrossRef]
33. Ilhan, B.; Yaman, H. Green Building Assessment Tool (GBAT) for Integrated BIM-Based Design Decisions. *Autom. Constr.* **2016**, *70*, 26–37. [CrossRef]
34. Marrero, M.; Wojtasiewicz, M.; Martínez-Rocamora, A.; Solís-Guzmán, J.; Alba-Rodríguez, M.D. BIM-LCA Integration for the Environmental Impact Assessment of the Urbanization Process. *Sustain.* **2020**, *12*, 4196. [CrossRef]
35. Sanhudo, L.; Martins, J.P. Building Information Modelling for an Automated Building Sustainability Assessment. *Civ. Eng. Environ. Syst.* **2018**, *35*, 99–116. [CrossRef]
36. Kamaruzzaman, S.N.; Salleh, H.; Weng Lou, E.C.; Edwards, R.; Wong, P.F. Assessment Schemes for Sustainability Design through BIM: Lessons Learnt. In Proceedings of the 4th International Building Control Conference 2016 (IBCC 2016), Kuala Lumpur, Malaysia, 7–8 March 2016.
37. Lazar, N.; Chithra, K. A Comprehensive Literature Review on Development of Building Sustainability Assessment Systems. *J. Build. Eng.* **2020**, *32*, 101450. [CrossRef]
38. Mateus, R.; Braganca, L. Sustainability Assessment and Rating of Buildings: Developing the Methodology SBToolPT–H. *Build. Environ.* **2011**, *46*, 1962–1971. [CrossRef]
39. Maltese, S.; Tagliabue, L.C.; Cecconi, F.R.; Pasini, D.; Manfren, M.; Ciribini, A.L. Sustainability Assessment through Green BIM for Environmental, Social and Economic Efficiency. *Procedia Eng.* **2017**, *180*, 520–530. [CrossRef]
40. Castro, M.F.; Mateus, R.; Braganca, L. Development of a Healthcare Building Sustainability Assessment Method—Proposed Structure and System of Weights for the Portuguese Context. *J. Clean. Prod.* **2017**, *148*, 555–570. [CrossRef]

41. Bragança, L. SBTool Urban: Instrumento para a Promoção da Sustentabilidade Urbana, in: I Simpósio Nacional de Gestão e Engenharia Urbana, São Carlos. Available online: http://civil.uminho.pt/urbenere/wp-content/uploads/2018/05/E28-SINGEURB-2017.pdf (accessed on 28 December 2018).
42. Roh, S.; Tae, S.; Suk, S.J.; Ford, G.; Shin, S. Development of a Building Life Cycle Carbon Emissions Assessment Program (BEGAS 2.0) for Korea's Green Building Index Certification System. *Renew. Sustain. Energy Rev.* **2016**, *53*, 954–965. [CrossRef]
43. Jrade, A.; Jalaei, F. Integrating Building Information Modelling with Sustainability to Design Building Projects at the Conceptual Stage. *Build. Simul.* **2013**, *6*, 429–444. [CrossRef]
44. Jalaei, F.; Jrade, A. Integrating Building Information Modeling (BIM) and LEED System at the Conceptual Design Stage of Sustainable Buildings. *Sustain. Cities Soc.* **2015**, *18*, 95–107. [CrossRef]
45. Wong, J.K.-W.; Kuan, K.-L. Implementing 'BEAM Plus' for BIM-Based Sustainability Analysis. *Autom. Constr.* **2014**, *44*, 163–175. [CrossRef]
46. Gandhi, S.; Jupp, J. BIM and Australian Green Star Building Certification. In *Computing in Civil and Building Engineering (2014), Proceedings of the 2014 International Conference on Computing in Civil and Building Engineering, Orlando, FL, USA, 23–25 June 2014*; American Society of Civil Engineers: Reston, VA, USA, 2014; pp. 275–282.
47. Kreiner, H.; Passer, A.; Wallbaum, H. A New Systemic Approach to Improve the Sustainability Performance of Office Buildings in the Early Design Stage. *Energy Build.* **2015**, *109*, 385–396. [CrossRef]
48. Alshamrani, O.S.; Galal, K.; Alkass, S. Integrated LCA–LEED Sustainability Assessment Model for Structure and Envelope Systems of School Buildings. *Build. Environ.* **2014**, *80*, 61–70. [CrossRef]
49. Muller, M.F.; Esmanioto, F.; Huber, N.; Loures, E.R.; Canciglieri, O.; Junior, O.C. A Systematic Literature Review of Interoperability in the Green Building Information Modeling Lifecycle. *J. Clean. Prod.* **2019**, *223*, 397–412. [CrossRef]

MDPI

Article

Development of Building Information Modeling Template for Environmental Impact Assessment

Sungwoo Lee [1], Sungho Tae [2,*], Hyungjae Jang [2,*], Chang U. Chae [3] and Youngjin Bok [4]

1 GHG Inventory Management Team, Greenhouse Gas Inventory and Research Center of Korea, Seoul 03181, Korea; greennaver@gmail.com
2 Department of Smart City Engineering, Hanyang University, Ansan 15644, Korea
3 Korea Institute of Civil Engineering and Building Technology, Goyang 10454, Korea; cuchae@kict.re.kr
4 GHG Reduction Team, Greenhouse Gas Inventory and Research Center of Korea, Seoul 03181, Korea; bokyoungjin@korea.kr
* Correspondence: jnb55@hanyang.ac.kr (S.T.); duethj@gmail.com (H.J.); Tel.: +82-31-400-5187 (S.T.); Fax: +82-31-406-711 (S.T. & H.J.)

Abstract: Eco-friendly building designs that use building information modeling (BIM) have become popular, and a variety of eco-friendly building assessment technologies that take advantage of BIM are being developed. However, existing building environmental performance assessment technologies that use BIM are linked to external assessment tools, and there exist compatibility issues among programs; it requires a considerable amount of time to address these problems, owing to the lack of experts who can operate the programs. This study aims to develop eco-friendly templates for assessing the embodied environmental impact of buildings using BIM authoring tools as part of the development of BIM-based building life cycle assessment (LCA) technologies. Therefore, an embodied environmental impact unit database was developed, for major building materials during production and operating stages, to perform embodied environmental impact assessments. Moreover, a major structural element library that uses the database was developed and a function was created to produce building environmental performance assessment results tables, making it possible to review the eco-friendliness of buildings. A case study analysis was performed to review the feasibility of the environmental performance assessment technologies. The results showed a less than 5% effective error rate in the assessment results that were obtained using the technology developed in this study compared with the assessment results based on the actual calculation and operating stage energy consumption figures, which proves the reliability of the proposed approach.

Keywords: building information modeling; building information modeling template; BIM library; life cycle assessment

Citation: Lee, S.; Tae, S.; Jang, H.; Chae, C.U.; Bok, Y. Development of Building Information Modeling Template for Environmental Impact Assessment. *Sustainability* **2021**, *13*, 3092. https://doi.org/10.3390/su13063092

Academic Editor: Antonio Garcia-Martinez

Received: 23 January 2021
Accepted: 2 March 2021
Published: 11 March 2021

1. Introduction

Global interest in environmental issues is increasing owing to severe environmental pollution worldwide. Thus, countries have been using the life cycle assessment (LCA) method proposed in the ISO 14000 series of international standards to perform environmental impact assessments across various fields of industry, while efforts are being made to reduce various kinds of environmental loads, including global warming [1]. In the construction industry, the LCA method is being introduced to assess the various environmental impacts caused by construction activities. Recently, to emphasize the importance of building environmental impact assessments, certification standards for LCAs were added to the 2016 Korean green building certification program (G-SEED), and the guidelines were revised and publicly announced [2]. A building LCA includes an assessment of the environmental impact of construction materials, regarded as the embodied environmental impact of the building. Internationally, research is being conducted to develop building

LCA programs and to make it possible to easily assess the embodied environmental impact of building materials [3].

However, to use BIM-based building embodied environment impact assessment technology in a more practical way, the following improvements must be made. First, it is necessary to conduct embodied environmental impact assessments that use a life cycle inventory (LCI) database, which differentiates construction materials based on their physical properties such as strength. Even when concrete is of the same type, there may be differences in environmental impact units based on the strength of the concrete, which has a significant effect on the results of building embodied environment impact assessments [4]. Second, the LCA certification standards in the G-SEED program must consider various categories of environmental impacts caused by a building, not just for the global warming potential. Therefore, in addition to global warming, which has been the subject of many studies in the past, other environmental impacts, including resource depletion potential [5] must be considered. Third, it is necessary to consider BIM-based studies that examine environmental cost assessments. Currently, various types of environmental impacts are being considered in terms of their indirect social costs so that they can be examined in monetary terms, known as the environmental cost, and the significance of this cost is increasing. Fourth, the environmental impact of energy consumption during the operation stage, which accounts for more than 60% of energy consumption during a building's entire lifespan, must be considered, starting from the design stage [6].

Therefore, the goal of this study is to create a BIM template with environmental impact parameters (BTEI), which can assess six categories of environmental impacts and environmental costs caused by specific building materials, as well as the design stage environmental impacts, to improve the accuracy of BIM-based building embodied environmental impact assessments.

To achieve this, environmental impact categories and environmental costs were defined. Six environmental impacts and environmental cost units were created for major building elements, and libraries were created for assessing the six environmental impacts as well as environmental costs. The libraries can be used in the BIM authoring tool Revit, and they consist of 3D objects with associated technical data. Revit allows for additional parameters to be created and applied to the libraries, and the data can be processed by inserting formulas for parameters and calculating the results.

This study used Autodesk Revit Architecture 2015 to model major structural elements, and it included database information that allows the environmental impact of building materials to be assessed.

KS I ISO 14025 presents a list of various environmental impact types that affect the global environment [7–9]. The environmental product declaration (EPD) certification system, which is operated by the Korean Ministry of Environment, assesses six categories of environmental impacts, including global warming (GWP), abiotic depletion (ADP), acidification potential (AP) eutrophication (EP), ozone layer depletion (ODP), and photochemical oxidation (POCP) [10]. The Korean Ministry of Environment has developed and released the LCI database in which substances that have an impact by categories are listed by product and material, and the Ministry distributes an LCA program to assess the environmental performance of products [11].

The LCI database can be calculated to develop environmental impact units based on the environmental impact categories using classification and characterization stages for each substance. These units can then be multiplied by a quantity based on the product's unit to calculate an environmental impact value.

This study used a library to develop a database for assessing the operating stage environmental impact of buildings, based on the designed building's total floor area.

Figure 1 shows the development method and assessment scope of the BTEI system proposed in this study. A template can be defined as an environment in which building environmental assessments can be performed autonomously in the BIM authoring tool without connecting to external assessment software. This study was performed using the Autodesk Revit Architecture 2015. To confirm the significance of this study, the environmental impact assessment results produced using the BIM library, considering the proposed environmental impact parameters, were compared with those based on actual calculation results and energy consumption.

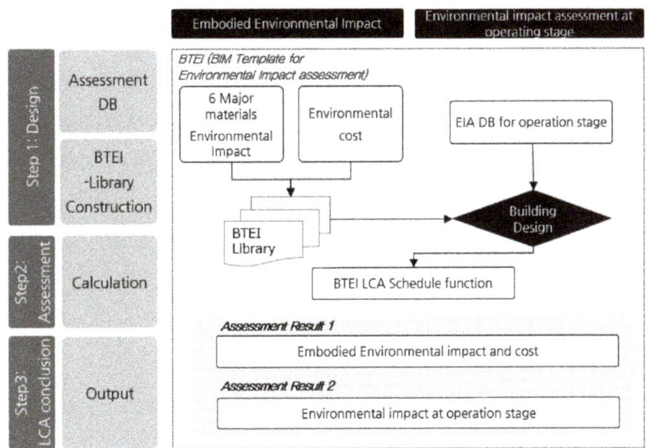

Figure 1. Development and scope of building information modeling (BIM) template with environmental impact parameters (BTEI).

2. BTEI Library Construction

2.1. Overview

The environmental impact assessment library created in this study is a BIM library that includes the environmental impact and environmental cost unit parameter data for major construction materials so that environmental impact and environmental cost assessments can be performed automatically for each material based on calculations [12]. In this study major construction materials were selected by considering the results of previous studies, and subdivided by the environmental information of materials to increase the library's accuracy [13]. The selected materials were used for library modeling. In addition, the national LCI database and the results of previous studies were used to analyze environmental impact and environmental cost units These database were recreated within the library as parameters to develop the library.

2.2. Selecting Major Construction Materials and Elements

Major construction materials were selected by considering the materials that cause environmental load among building construction materials, along with the construction material information that was calculated at a level of detail (LOD) of 300, which is the modeling standard suggested by the Korean Ministry of Land, Infrastructure, and Transport's BIM guidelines. By referring to previous studies that analyzed materials accounting for 95% of the environmental impact of materials used in buildings, concrete, rebar, steel framing, glass, insulation, gypsum board, and concrete products were initially selected in this study [14].

However, instead of concrete products, bricks were selected as an assessment target, as they are predominantly used in the construction industry in South Korea. However, rebar and steel framing were excluded as they were not a part of the LOD 300 standard; five major construction materials (concrete, glass, bricks, insulation, gypsum board, and concrete products) were selected [15].

In addition, the library was constructed with six structural elements (walls, columns, beams, slabs, windows, and foundations) as suggested by Korea's Ministry of Land, Infrastructure and Transport.

2.3. Calculating Environmental Impact Parameters

To create environmental impact parameters in the assessment library, the LCI database entries for each construction material were analyzed, and environmental impact units and environmental cost units were calculated, as shown in Table 1 [16].

As observed in the ready-mixed concrete item in Table 2, different unit values were calculated for the same construction material depending on the specific type [17]. In the case of ready-mixed concrete, it was found that there were differences in the LCI database based on the ratio of cement and admixture. Because of this, there were differences in the environmental impact unit based on the strength of the ready-mixed concrete.

Parameters were created in the library that models these six environmental impact units and environmental cost units, and a BTEI library containing these environmental impact parameters was developed.

In the component library, elements were modeled as shown in Figure 2 and saved in separate files. Similarly, parameters were added via type characteristics and saved in the environmental impact assessment unit database. After the library was developed, it was linked to the assessment unit database, and the Revit Assessment table was set up to allow automatic calculations. Thus, it was possible to monitor the assessment results when users performed modeling using the BTEI library [18].

2.4. Development of BIM Library That Considers Environmental Impact Parameters

The BTEI library comprises six elements with five major construction materials, including 12 walls, 6 columns, 30 windows, 6 beams, 6 slabs, and 12 foundations. Of these, the window, column, and foundation elements are system libraries that were constructed as separate files, and the wall, slab, and beam elements are component libraries that were constructed within a task file.

The wall and slab libraries were constructed as single construction materials so that the user could model combined elements based on the project. In addition, owing to the variety of modern construction projects and the increasing height of apartment buildings, concrete of varying strength and various types of insulation, glass, etc. are used for each part of a building. Therefore, to facilitate environmental impact assessments that consider this variety, libraries were developed to account for the specific characteristics of materials, as shown in Table 2 [19].

Wall elements use major construction materials such as concrete, gypsum board, and insulation. Six concrete wall elements were created to take into account concrete of varying strength (21, 24, 27, 30, 35, and 40 MPa) [20].

Similarly, elements such as beams, slabs, and foundations, made of concrete, were created considering the six varying concrete strengths. Among the construction materials, concrete is a major emissions source of the GWP, and its environmental impact units are calculated differently owing to the difference in the LCI database for each concrete strength. Therefore, when assessing the embodied environmental impact of a building, the use of environmental impact units for each concrete strength improves the accuracy of the building LCA results. Consequently, in the BIM-based LCAs using the BTEI library created in this study, the accuracy of the assessment results can be improved by using environmental impact units for each concrete strength [21].

In the case of the BTEI-library-wall items, an LCI database with three different types of insulation (glass wool, expanded polystyrene, and extruded polystyrene) was used to create three insulation walls. LCI database entries for concrete bricks and clay bricks were used to create two wall-brick items. Similarly, LCI database entries for three types of glass (plate, double glazing, and tempered) were used for each of 10 basic windows, such as four-unit sliding windows, single swinging windows, and fixed windows, to create 30 window libraries [22].

Table 1. Database for evaluating environmental impact parameters.

Major. Materials	Subdivision	Unit	6 Environmental Impact Database						Environmental Cost (Won)
			GWP [kg-CO_2eq/Unit]	ADP [kg/Unit]	AP [kg-SO_2eq/Unit]	EP [kg-PO_4^{3-} eq/Unit]	ODP [kg-CFC-11eq/Unit]	POCP [kg-Ethyleneeq/Unit]	
Ready-mixed Concrete	21 MPa	m^3	4.19×10^2	1.56×10^0	6.94×10^{-1}	8.08×10^{-2}	4.61×10^{-5}	1.13×10^0	1.53×10^4
	24 MPa	m^3	4.29×10^2	2.08×10^0	7.05×10^{-1}	8.20×10^{-2}	4.59×10^{-5}	1.15×10^0	1.56×10^4
Glass	Plate Glass	Ton	7.88×10^2	6.97×10^0	3.67×10^0	5.23×10^{-2}	3.04×10^{-4}	8.95×10^{-1}	5.60×10^4
Brick	Concrete Brick	1000 EA	1.23×10^{-1}	1.46×10^{-4}	1.57×10^{-4}	2.27×10^{-5}	4.71×10^{-9}	1.31×10^{-5}	2.90×10^0
Insulation	Glass Wool	Kg	1.05×10^0	5.06×10^{-1}	1.83×10^1	4.95×10^{-1}	5.02×10^{-1}	7.39×10^{-4}	2.68×10^5
Gypsum Board	-	Ton	1.92×10^{-1}	1.55×10^{-2}	3.13×10^{-2}	5.28×10^{-3}	5.67×10^{-7}	7.61×10^{-3}	4.47×10^2

GWP: Global warming; ADP: Abiotic depletion; AP: Acidification potential; EP: Eutrophi-cation; ODP: Ozone layer depletion; POCP: Photochemical oxidation.

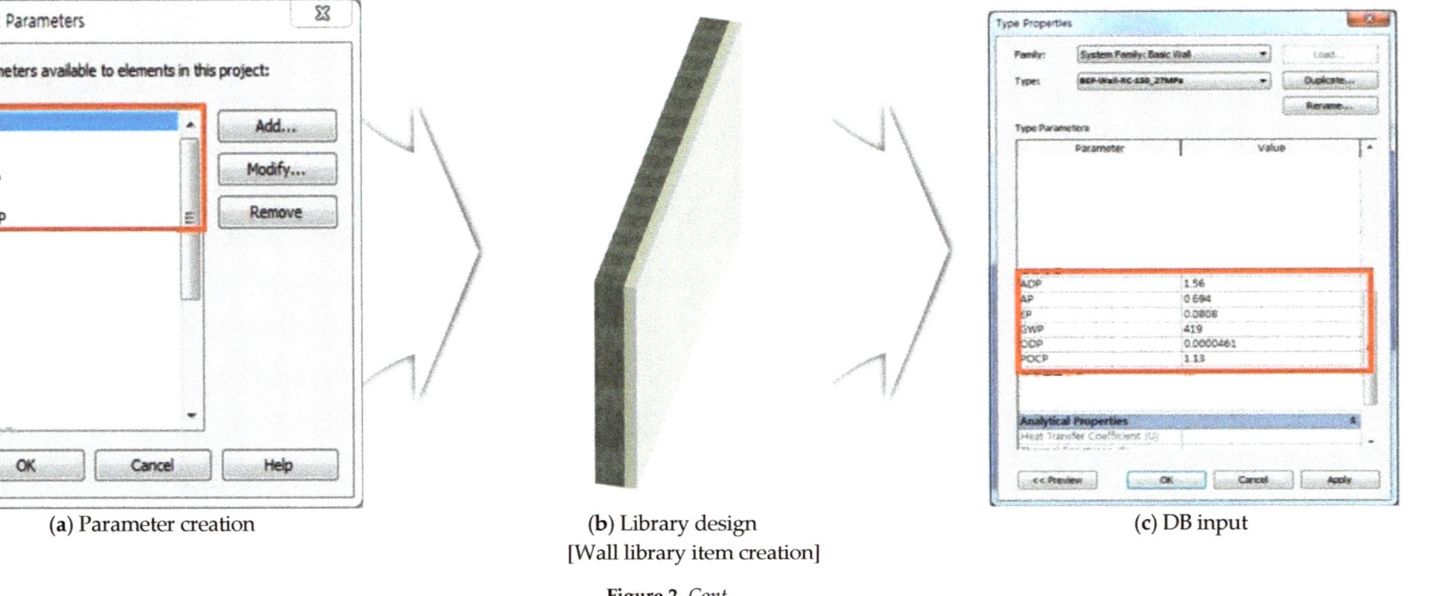

(**a**) Parameter creation (**b**) Library design [Wall library item creation] (**c**) DB input

Figure 2. *Cont.*

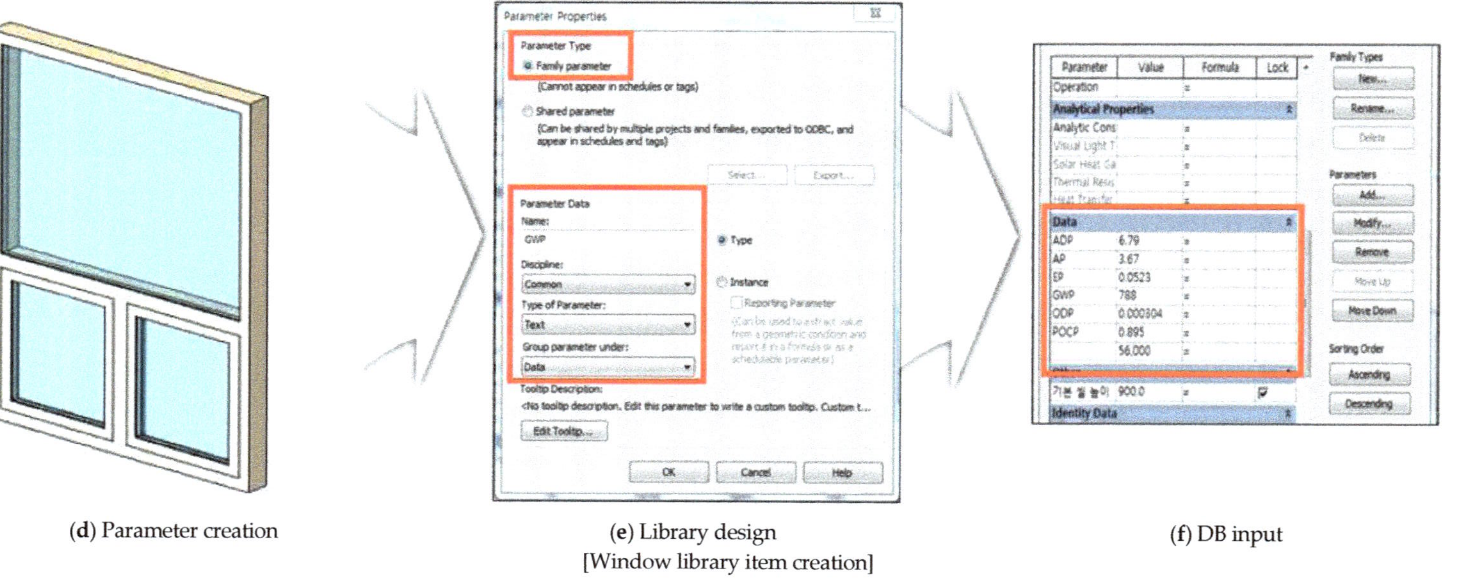

(**d**) Parameter creation

(**e**) Library design [Window library item creation]

(**f**) DB input

Figure 2. BTEI-library development method.

Table 2. Subdivision in materials of each elements.

Elements	Material	Subdivision	File Format
Wall	Ready-mixed concrete	Strength	Inplace Library
	Insulation	Type	Inplace Library
	Brick	Type	Inplace Library
	Gypsum board	-	Inplace Library
Slab	Ready-mixed concrete	Strength	Inplace Library
Beam	Ready-mixed concrete	Strength	Inplace Library
Window	Glass	Type	System Library
Column	Ready-mixed concrete	Strength	System Library
Foundation	Ready-mixed concrete	Strength	System Library

The environmental impact and environmental cost assessment results were found for three environmental impact types (GWP, AP, and ODP) of several libraries of given sizes (BTEI-library-wall 200 mm, BTEI-library-column 300 × 300 × 2700 mm^3, BTEI-slab 150 mm, BTEI-window-single-swinging) among the libraries that were created, by subdividing the building materials, as shown in Table 3.

Table 3. BTEI-library construction example.

Elements	Library Name	Subdivision			Environmental Cost (won)
		GWP [$kgCO_2eq$/Unit]	AP [$kgSO_2eq$/Unit]	ODP [kg-CFC-11eq/Unit]	
Wall-RC 6 species	BTEI-Wall-RC_21MPa	8.38×10^0	1.39×10^{-2}	9.22×10^{-7}	3.07×10^2
	BTEI-Wall-RC_24MPa	8.58×10^0	1.41×10^{-2}	9.18×10^{-7}	3.12×10^2
	BTEI-Wall-RC_27MPa	4.79×10^0	7.42×10^{-4}	5.27×10^{-6}	8.05×10^1
Column-RC 6 species	BTEI-Column-RC_21MPa	4.53×10^2	7.50×10^{-1}	4.98×10^{-5}	1.66×10^4
	BTEI-Column-RC_24MPa	4.63×10^2	7.61×10^{-1}	4.96×10^{-5}	1.69×10^4
	BTEI-Column-RC_27MPa	2.59×10^2	4.01×10^{-1}	2.85×10^{-4}	9.07×10^3
Slab-RC 6 species	BTEI-Slab-RC_21MPa	6.29×10^0	1.04×10^{-2}	6.92×10^{-7}	2.30×10^2
	BTEI-Slab-RC_24MPa	6.44×10^0	1.06×10^{-2}	6.89×10^{-7}	2.34×10^2
	BTEI-Slab-RC_27MPa	3.60×10^0	5.57×10^{-4}	3.95×10^{-6}	6.04×10^1
Window 30 species	BTEI-Window-Single Swing-Plate Glass	7.88×10^2	3.67×10^0	3.04×10^{-4}	5.60×10^4
	BTEI-Window-Single Swing-Double Glazing	2.24×10^2	3.05×10^{-1}	1.81×10^{-6}	7.09×10^3
	BTEI-Window-Single Swing-Tempered Glass	1.34×10^2	2.57×10^{-1}	6.64×10^{-7}	4.74×10^3

In the case of the BTEI-Wall-RC-200 libraries, which are developed for concrete with the same thickness, the environmental impact and environmental cost values were calculated differently, based on strength. As mentioned before, there may be a difference in the environmental impact and environmental cost units according to the subdivided material information, even among libraries that are composed of the same construction material.

Therefore, the results of assessments that use these libraries may also differ, as shown in Table 3.

Libraries were developed so that the users could set up elements of all sizes, and model the target elements by combining single material libraries. By doing so, modeling could be performed in response to a variety of construction projects, and environmental impact assessments could be performed on the materials used in buildings.

3. Environmental Impact Assessments on the Operation Stage Using BTEI

3.1. Outline

Generally, energy and maintenance costs have an important impact on decision-making during building planning and the basic design stage in LCAs. This study aims to create a system that can take into account the environmental impacts of the material production stage and operating stage energy consumption in building LCAs [23].

To achieve this, an environmental impact assessment database was developed in this study for each energy source used in a building. To consider the design stage energy consumption, 12 energy consumption databases for various building uses were provided to examine the operation stage environmental impact of buildings designed with the BTEI [24].

3.2. Development of Environmental Impact Assessment Database for the Operation Stage

The environmental impacts of operating stage energy sources occur during both the production stage and the combustion stage. Therefore, to calculate the environmental impact units of energy sources, the environmental impact of the energy source's production stage and combustion stage must be considered simultaneously, as shown in Equation (1). This study selected the national LCI database as the database for energy source production stages, and calculated the environmental impact units using the same method as the construction material (classification and specialization of the LCI database). However, the LCI database for the energy source production stage was not developed yet as a national LCI database, so the results of previous environmental impact analyses were used in this study. Table 4 shows the environmental impact units for the energy sources.

$$EIC_{i,j} = \sum_{k}\left(PE_{i,k} \times IF_{j,k} + CE_{i,k} \times IF_{j,k}\right) \quad (1)$$

Table 4. Database of environmental impact assessments by energy source.

Energy Source	Fu	Stage	Environmental Impact Category					
			GWP	AP	EP	ODP	POCP	ADP
Electricity	kWh	Total	4.88×10^{-1}	8.37×10^{-4}	1.56×10^{-4}	1.37×10^{-11}	1.41×10^{-6}	8.58×10^{-4}
Heat	Gcal	Production	3.20×10^{-1}	2.96×10^{-3}	3.39×10^{-4}	1.30×10^{-6}	7.50×10^{-3}	2.34×10^{-2}
		Combustion	2.87×10^{0}	7.90×10^{-3}	1.40×10^{-3}	-	3.74×10^{-3}	-
		Total	3.19×10^{0}	1.09×10^{-2}	1.74×10^{-3}	1.30×10^{-6}	1.12×10^{-2}	2.34×10^{-2}
Kerosene	ℓ	Production	8.29×10^{-2}	1.86×10^{-4}	1.07×10^{-5}	2.61×10^{-10}	8.28×10^{-6}	2.18×10^{-2}
		Combustion	2.19×10^{0}	4.34×10^{-3}	7.63×10^{-4}	-	7.14×10^{-3}	-
		Total	2.27×10^{0}	4.53×10^{-3}	7.74×10^{-4}	2.61×10^{-10}	7.15×10^{-3}	2.18×10^{-2}
City gas	m^3	Production	4.96×10^{-1}	2.77×10^{-3}	1.13×10^{-4}	4.24×10^{-9}	1.87×10^{-2}	2.16×10^{-2}
		Combustion	3.11×10^{0}	3.37×10^{-3}	6.23×10^{-4}	-	1.98×10^{-4}	-
		Total	3.61×10^{0}	6.14×10^{-3}	7.36×10^{-4}	4.24×10^{-9}	1.89×10^{-2}	2.16×10^{-2}

The unit of environmental impact categories, "GWP" is the kg-CO_{2eq}/FU; "AP" is the kg-SO_{2eq}/FU; "EP" is the kg-$PO_4{}^{3-}{}_{eq}$/FU; "ODP" is the kg-CFC-11_{eq}/FU; "POCP" is the kg-C_2H_{4eq}/FU; "ADP" is the kg-Sb_{eq}/FU.

Here, $EIC_{i,j}$ is the environmental impact unit of environmental impact type *(j)* for the function unit of energy source *(i)*. $PE_{i,k}$ is the emissions of the impact substance *(k)* for each function unit caused by the production of energy source *(i)*. $IF_{j,k}$ is the impact factor of impact substance *(k)* for the environmental impact type *(j)*. $CE_{i,k}$ is the emissions of impact substance *(k)* for each function unit caused by the combustion of energy source *(i)* [25].

3.3. Development of Building Energy Consumption Unit Database

In this study, the amount of energy consumed for each usage type is provided to perform operating stage environmental impact assessments on buildings designed by the BTEI. Previous studies have proposed managing the environmental performance of buildings by including the buildings' usage types as well as the occupants in a database, if they are apparent [26].

To this end, the report of the Korean Energy Census was used as basic data, and the energy consumption was predicted by dividing data into household and commercial sectors.

In addition, the basic unit of evaluation based on the total floor area was established using the total floor area data for residential and commercial areas provided by the Ministry of Land, Infrastructure, and Transport. Analysis was performed on the energy of buildings with limited building functions and usage types, in particular the energy consumption caused by the buildings' heating/cooling operation, and these data were developed in units [27].

Table 5 shows the data types used as basic data for understanding energy consumption here. Basic data were collected from the Korea Building Energy Integrated Management System for 1398 buildings in Seoul, including office (795), education (16), medical (25), research (15), recycling (29), neighborhood living (356), training (9), cultural (83), automotive (11), sports (12), storage (7), and tourism facilities (32). The basic data included information on the building structure, building floor area, heating/cooling style, energy sources, and energy consumption [28].

Table 5. Categories of basic data for constructing energy statistics.

Information	Data
Energy	Electricity, Heat, Kerosens, City Gas
Heating/Cooling Style	Boiler and heat circulation pump, individual heating equipment, central heat source and individual heating equipment thermal production
	Central cooling/heating source (refrigeration equipment, cooling tower, cool water circulation pump) and cooling circulation pump, individual cooling equipment (EHP, GHP, PAC, etc.) Includes operational electricity consumption other than cooling/heating production of central cooling/heating source and individual cooling equipment
Basic Information	Building floor area

The basic data were used to provide consumption statistics for each of the buildings' energy sources to enable the prediction of the energy consumption resulting from the buildings' floor area as designed by the user.

Energy statistics for each building usage type were used, as shown in Table 6, to assess the buildings' environmental impact in concrete terms and to consider their environmental impact during the design stage.

Table 6. Energy statistics for each building use.

Classification	Information	Electricity (KWh/m^2)	Heat (Gcal/m^2)	Kerosens (ℓ/m^2)	City Gas (m^3/m^2)
Residential	Detached house	7.407×10^{-2}	-	5.533×10^{-6}	7.667×10^{-6}
	Apartment	2.856×10^{-2}	1.780×10^{-2}	4.368×10^{-8}	5.167×10^{-6}
	Townhouse	1.091×10^{-1}	9.561×10^{-4}	5.777×10^{-7}	2.477×10^{-5}
	Multiplex housing	4.315×10^{-2}	-	1.126×10^{-7}	9.813×10^{-6}
Commercial	Retail business	4.699×10^{-1}	6.331×10^{-4}	5.757×10^{-7}	6.356×10^{-6}
	Lodging business	1.789×10^{-1}	3.160×10^{-4}	4.225×10^{-7}	7.601×10^{-6}

The energy statistics were used to develop a unit factor per floor area. The system makes it possible to use the floor area information extracted from the BTEI, to predict energy consumption, and the environmental impact assessment database to assess the operation stage.

An example design was made to estimate the energy consumption for each building function within a fixed level of allowable error. Therefore, statistical values representing the unit distribution characteristics were used as mode values [29].

The units for 12 target building functions were compared, and the results showed that recycling facilities showed the biggest units, followed by storage facilities and sports facilities. Because these facilities have a fixed consumption demand, their energy saving potential was considered to be lower than other buildings. However, office, education, and medical facilities, which have long lifetimes and relatively high units, were considered to have higher energy saving potential than the others.

4. Case Study

4.1. Outline

According to ISO 21931-1, a building's life cycle stages can be divided into the production, construction, operating, and disposal stages as shown Figure 3. The BTEI system that was developed in this study targets the production and operating stages. Therefore, this study used scenarios for each stage, as shown in Figure 4, to perform life cycle environmental impact assessments [30,31].

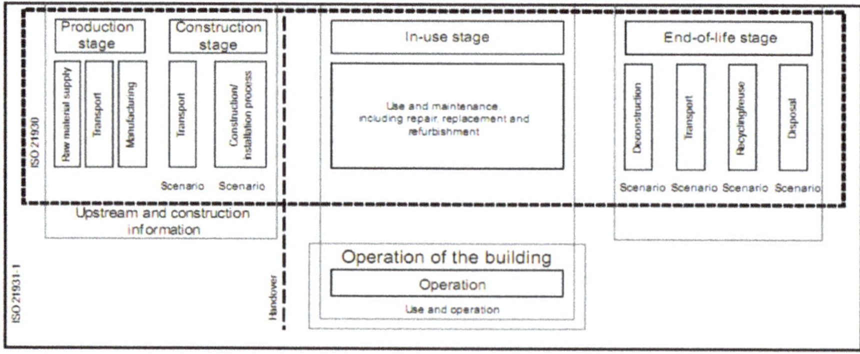

Figure 3. Life cycle assessment (LCA) scope according to ISO 21931-1.

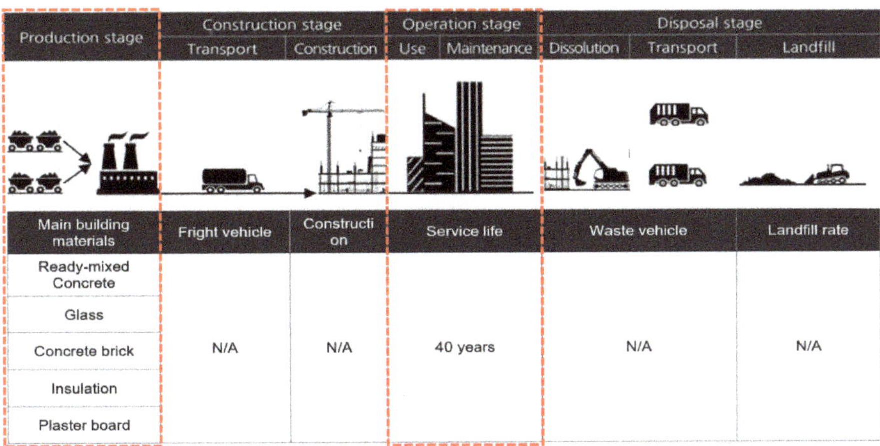

Figure 4. BTEI scenario by LCA stage.

This study calculated the major construction materials for the building production stage, which cause at least 95% of the major environmental impact types (GWP, AP, EP, ODP, POCP, and ADP) based on the environmental impact assessment results for buildings by previous studies.

This is because, in the production stage, construction materials account for at least 90% of the embodied environmental impact that occurs during the life cycle stages of the building [29].

In the production stage, the libraries provided by the BTEI environment were used and designed, and the embodied environmental impact during the buildings' life cycle stages was assessed [30].

To assess the environmental impact occurred in the operation stage, it is necessary to set the building's lifespan. The lifespan of a building could be subdivided based on its purpose and certain criteria in terms of its physical, functional, societal, economical, and legal lifespans. To assess the environmental impact of buildings in the initial stages of their construction and compare the assessment results, this study used a fixed lifespan of 40 years, which is stipulated by the Korean Corporate Tax Act for buildings [31].

4.2. Assessment Method

Verification was performed on the six environmental impact and environmental cost calculations, performed in BIM using the proposed BTEI system. In addition, the environmental impact caused by energy consumption was simulated using the operating stage environmental impact assessment database and energy statistics.

Environmental impact assessments were performed on the operation stages and the materials used in the actual building, as shown in Table 7. The assessment target was the, which is an apartment housing in Busan of South Korea. It is a residential building with 18 stories, made of reinforced concrete. Major construction materials with various characteristics were used in the building.

Table 7. Overview of evaluation target.

Information	
Project	An Apartment in Busan
Purpose	Apartment
Structural	Reinforced concrete
Scale	18 floors
Life expectancy	40 years
Total ground areas	5313.90 m^2
Floor area	282.95 m^2

27-MPa concrete was used from the 1st floor to the 7th floor, and 24-MPa concrete was used from the 8th floor to the 18th floor. Expanded polystyrene was used as the insulation, and double glazing glass and plate glass were used for the windows.

Environmental impact assessments were performed using the BTEI-based modeling and also with manual calculation, and the assessments were analyzed. First, the target building's actual calculation results were used to calculate the quantities of major construction materials such as concrete, glass, insulation, gypsum board, and bricks to directly assess the embodied environmental impact and environmental cost of each material.

Autodesk Revit Architecture 2015 was used to apply the BTEI, and modeling was performed on the assessment target at an LOD of 300. The BTEI modeled the concrete, bricks, glass, insulation, gypsum board, and concrete products and calculated the assessment results.

Modeling was performed using the BTEI library, which includes the environmental impact units, environmental cost units, and quantity unit conversion factor in the BIM task file. The BTEI-environmental impact assessment table function was used to display the results. Here, the unit conversion factor was a parameter needed to convert the volume unit quantities calculated in BIM to the standard units of the environmental impact units. Table 8 shows the unit conversion factor Weights were used for each construction material to calculate the unit conversion factor [32].

Table 8. Unit conversion factor.

Major Materials	Revit S/W Unit	6 Environmental Impacts Unit	Unit Conversion Factor
Remicon	m^3	m^3	1
Plate Glass	m^3	ton	0.0119
Double Glazing	m^3	m^2	200
Concrete Brick	m^3	1000 ea	0.75
Clay Brick	m^3	ton	192
Insulation-Foam	m^3	ton	0.03
Insulation-EPS	m^3	kg	0.16
Gypsum Board	m^3	ton	0.863

With regard to the operation stage, the environmental impact assessment results for 40 years, of the energy usage in the actual building and the building floor area that was modeled in the BTEI were used to analyze the assessment results that employed the operating stage environmental impact assessment units [33].

4.3. Results of Environmental Impact Assessment on Major Building Materials

4.3.1. Environmental Impact Assessment Results

A comparison of the assessment results was performed through a percentage analysis based on the manual calculation results, to compare the assessment results considering the six environmental impacts and environmental costs that were calculated using actual data and calculation by the BIM authoring tool (using the BTEI for each of the six environmental impact types).

Figure 5a shows the assessment results for each environmental impact type. In the BTEI, the overall results were similar to the actual calculation based assessment results, the error being just below 0.01%. It was found that the error in acidification (AP), which had a 15% error compared to the ozone layer depletion (ODP) parameter, was caused by the type of insulation material used in the previously analyzed materials' environmental impact units. That is, it was found that the error was caused by using foam insulation, which was the only insulation library created in the BTEI, rather than using the expanded polystyrene that was used in the actual building. Figure 5b demonstrates the material production stage environmental cost assessment results. In Figure 5b, there was a difference of approximately 4% between the BTEI's environmental cost assessment results and the actual calculation based assessment results. Table 9 shows the environmental impact and environmental cost assessment results for each material when using the supply calculation records and the BTEI on an apartment building. It was found that the BTEI assessment results were similar to those based on the supply calculation records for all six environmental impacts and environmental costs.

The global warming assessment results for concrete, which accounts for a large portion of the building's embodied environmental impact, exhibited an error rate of approximately 1.3%, considering the environmental impact units for each strength (24 and 27 MPa).

It was determined that the BTEI results were similar to the actual supply calculation records because its libraries were constructed using environmental data that were subdivided for each construction material, confirming the significance of this study.

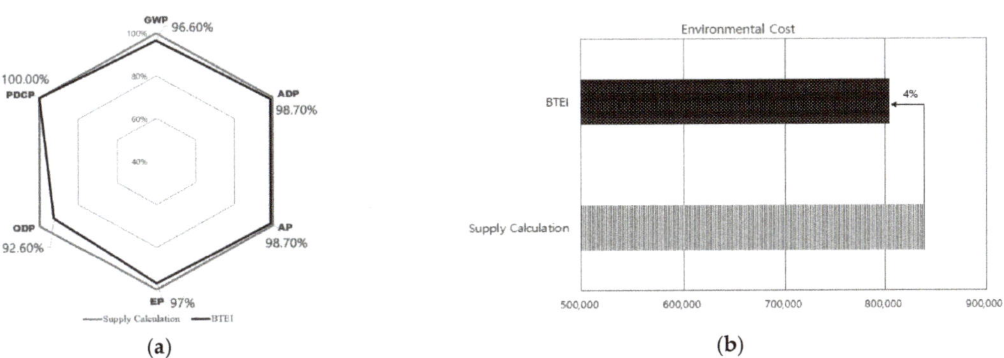

(a) (b)

Figure 5. Analysis results. (**a**) Relative analysis of environmental impact between supply calculation and the BTEI. (**b**) Relative analysis of environmental cost between supply calculation and the BTEI.

Table 9. Environmental impact and environmental cost for each major material.

Category	Material	6 Environmental Impact						Environmental Cost
		GWP	ADP	AP	EP	ODP	POCP	
BTEI (BIM Template for Environmental Impact Evaluation)	Remicon 24 MPa	2.12×10^{2}	1.03×10^{0}	3.48×10^{-1}	4.04×10^{-2}	2.26×10^{-5}	5.67×10^{-1}	7.70×10^{3}
	Remicon 27 MPa	6.67×10^{1}	2.06×10^{-1}	1.03×10^{-1}	1.21×10^{-2}	7.31×10^{-5}	1.75×10^{-1}	2.34×10^{3}
	Plate Glass	1.11×10^{0}	9.82×10^{-3}	5.17×10^{-3}	7.37×10^{-5}	4.28×10^{-7}	1.24×10^{-3}	7.88×10^{1}
	Double Glazing	5.30×10^{0}	2.16×10^{-2}	7.22×10^{-3}	5.23×10^{-4}	4.23×10^{-8}	1.26×10^{-2}	1.68×10^{2}
	Concrete Brick	2.02×10^{1}	2.40×10^{-2}	2.58×10^{-2}	3.73×10^{-3}	7.72×10^{-7}	2.13×10^{-3}	4.79×10^{2}
	EPS	3.66×10^{-2}	2.58×10^{-2}	1.80×10^{-1}	4.77×10^{-3}	2.29×10^{-2}	2.14×10^{-6}	3.92×10^{3}
	Gypsum Board	1.79×10^{-1}	1.45×10^{-2}	2.92×10^{-2}	4.93×10^{-3}	5.25×10^{-7}	7.09×10^{-3}	4.18×10^{2}
Supply Calculation	Remicon 24 MPa	2.09×10^{2}	1.02×10^{0}	3.44×10^{-1}	4.00×10^{-2}	2.21×10^{-5}	5.61×10^{-1}	7.62×10^{3}
	Remicon 27 MPa	6.51×10^{1}	2.01×10^{-1}	1.01×10^{-1}	1.18×10^{-2}	7.12×10^{-5}	1.71×10^{-1}	2.28×10^{3}
	Plate Glass	1.44×10^{0}	1.27×10^{-2}	6.70×10^{-3}	9.54×10^{-5}	5.52×10^{-7}	1.62×10^{-3}	1.02×10^{2}
	Double Glazing	8.77×10^{0}	3.57×10^{-2}	1.19×10^{-2}	8.65×10^{-4}	7.02×10^{-8}	2.10×10^{-2}	2.77×10^{2}
	Concrete Brick	3.09×10^{-2}	3.67×10^{-2}	3.95×10^{-2}	5.71×10^{-3}	1.12×10^{-6}	3.26×10^{-3}	7.33×10^{2}
	EPS	3.95×10^{-2}	2.79×10^{-2}	1.94×10^{-1}	5.15×10^{-3}	3.12×10^{-2}	2.39×10^{-6}	4.24×10^{3}
	Gypsum Board	1.82×10^{-1}	1.47×10^{-2}	2.96×10^{-2}	4.99×10^{-3}	5.24×10^{-7}	7.16×10^{-3}	4.23×10^{2}

GWP (kg-CO_2eq/unit), ADP (kg/Unit), AP (kg-SO_2eq/Unit), EP (kg-PO_4^{3-}eq/unit), ODP (kg-CFC-11eq/Unit), POCP (kg-Ethyleneeq/Unit), Environmental Cost (Won/m^2).

4.3.2. Results of Environmental Impact Assessment on the Operation Stage

In the BTEI Template, the operating stage environmental impact assessment database proposed in this study and the energy consumption statistical data for each building were added to the BTEI. The user could simulate the environmental impact caused by operating stage energy consumption using only a BIM design with an LOD 300.

This study compared and analyzed the environmental impact assessment results based on the case-study assessment target building's actual energy consumption and those calculated using the floor area of the target buildings as designed in the BTEI at an LOD 300. The assessment results are shown in Table 10.

Table 10. Operating stage environmental impact assessment results.

Category		Case Targeted Building	BTEI	Difference Rate
Energy Use	Electricity use (Kwh)	60,108,040	57,523,394	4.3%
Environmental Impact	GWP (Kg-CO_2eq/unit)	2.93×10^{7}	2.81×10^{7}	
	ADP (kg/Unit)	5.03×10^{4}	4.81×10^{4}	
	AP (kg-SO_2eq/Unit)	9.38×10^{3}	8.97×10^{3}	
	EP (Kg-PO_4^{3-}eq/unit)	8.23×10^{-4}	7.88×10^{-4}	
	ODP (kg-CFC-11eq/Unit)	8.48×10^{1}	8.11×10^{1}	
	PDCP (kg-Ethyleneeq/Unit)	5.16×10^{4}	4.94×10^{4}	

The building's life span was set to 40 years. For the case study, the actual average electricity consumption for the last five years was assumed for 40 years to calculate the assessment results. For the BTEI, the floor area-based energy consumption statistics were used to calculate the energy consumption, and the environmental impact assessment database was used to calculate the results. As shown in Table 10, there was a 4.3% difference between the BTEI assessment results and those based on the actual energy consumption, which confirms the significance of the proposed methodology.

5. Discussion

The purpose of this study was to create the BIM BTEI template, providing a method to assess the embodied environmental impact and operating stage environmental impact of buildings using the BIM authoring tool Revit as part of the development of assessment technologies for six categories of building environmental impacts.

A BIM template is a consistent input with a fixed, preset structure that is often used to obtain a calculated output suitable for certain assessment goals and scopes regarding the use of BIM modeling information [33].

Currently, studies on using BIM to assess the environmental performance of buildings are being conducted, but this approach is limited by the fact that the data lack of compatibility, no standard data, and the assessments can take too long [34,35].

With the BTEI approach, the expected results can be obtained rapidly using data that were previously provided by the user via templates. The BTEI provides major construction element libraries and tables for assessing the embodied environmental impacts of buildings in Revit. The major construction elements have been divided into walls, columns, slabs, and windows, and 46 libraries were created—the major construction materials that make up the elements being ready-mixed concrete, glass, concrete bricks, insulation, and gypsum board.

It is possible to simulate the operating stage energy consumption based on the building's designed floor area data and to use this to derive the six environmental impact assessment results.

The BTEI assesses the environmental impact of the material production stage and the building's operation stage, and it is possible to account for approximately 90% of the building's LCA. In addition, the system's biggest advantage is that various types of environmental performance can be considered without linking to other programs by using building information that can be extracted from the BIM system.

Construction materials such as rebars that cannot be considered in the Revit architecture may lower the reliability of the assessment results. Rebar is an element that must be considered in a building's embodied environmental impact assessment because it is a major construction material accounting for 20% or more of a building's embodied environmental impact. To improve the BTEI standards, it is necessary to allow rebar and premium rates for major construction materials to be considered for each building modeling LOD.

In this study, an assessment database was developed to consider the environmental impacts of the operation stage. However, it does not include various eco-friendly technologies and building materials which are currently being used to reduce the environmental impacts from the operation stage of buildings.

To improve the usefulness and reliability of the BTEI, it will be necessary to perform further studies that consider eight types of major building materials, including rebar and steel framing, rather than six types as well as studies that comprehensively assess the environmental impact of buildings that use specific materials and sustainable technologies to reduce environmental loads.

6. Conclusions

The goal of this study was to develop the BIM Template to evaluate the environmental impact of the building material production stage and operation stage using the Revit application, which is a BIM authoring tool, as part of research on the development of BIM-based building life cycle environmental impact assessment. The BIM template BTEI was developed, which made it possible to assess embodied and operational environmental impacts with environmental cost within a BIM authoring tool. When evaluating the sustainability of a building in the early stages of the construction project, the BIM Template developed in this study is very useful. If the accurate information on the input amount of building materials for an apartment house is insufficient, this research method is applied to evaluate the sustainability of the entire process of the building. This is because it is possible to make a comprehensive judgment. This study attempted to expand the area of analysis by incorporating the method of using BIM in the field of sustainability assessment throughout the entire building process. By presenting the environmental performance and economic evaluation method at the building design stage according to the characteristics of the building according to the characteristics of the building, the information on the input amount of building materials, which is a key area for the sustainability assessment at the initial stage of the building project, will be helpful in future studies on sustainability assessment throughout the entire building process. It is believed to be possible. In addition, 72 libraries including 6 categories of environmental impact and economics DB built in this study can be used as a systematic evaluation method of buildings, and furthermore, effectively support stakeholder decision-making to enhance the sustainability of the entire process. It is believed to be possible.

Author Contributions: Conceptualization, S.L., S.T. and H.J.; methodology, S.L. and S.T.; formal analysis, Y.B.; investigation, Y.B.; resources, S.L.; data curation, S.L. and Y.B.; writing—original draft preparation, S.L.; writing—review and editing, C.U.C.; visualization, S.L.; supervision, S.T.; funding acquisition, S.T. All authors have read and agreed to the published version of the manuscript.

Funding: This research was supported by Basic Science Research Program through the National Research Foundation of Korea (NRF) funded by the Ministry of Education, Science and Technology (No. 2018R1D1A1A09083678) and supported by the National Research Foundation of Korea (NRF) grant funded by the Korea government (MSIT) (No. 2015R1A5A1037548).

Institutional Review Board Statement: Not applicable.

Informed Consent Statement: Informed consent was obtained from all subjects involved in the study.

Data Availability Statement: Data sharing not applicable.

Conflicts of Interest: The authors declare no conflict of interest.

References

1. Cabeza, L.; Rincon, L.; Vilarinob, V.; Pereza, G.; Castell, A. Life cycle assessment (LCA) and life cycle energy analysis (LCEA) of buildings and the building sector: A review. *Renew. Sustain. Energy Rev.* **2014**, *29*, 394–416. [CrossRef]
2. Ministry of Land Infrastructure and Transport. *Green Building Certification System in Korea*; Ministry of Land Infrastructure and Transport: Sejong, Korea, 2013.
3. Roh, S.; Tae, S.; Shin, S. A Development of Building materials embodied greenhouse gases assessment criteria and system (BEGAS) in the newly revised Korea Green Building Certification System(G-SEED). *Renew. Sustain. Energy Rev.* **2014**, *35*, 410–421. [CrossRef]
4. Li, X.; Yang, F.; Zhu, Y.; Gao, Y. An assessment framework for analyzing the embodied carbon impacts of residential buildings in China. *Energy Build.* **2014**, *85*, 400–409. [CrossRef]
5. Seinre, E.; Kurnitski, J.; Voll, H. Building sustainability objective assessment in Estonian context and a comparative evaluation with LEED and BREEAM. *Build. Environ.* **2014**, *82*, 110–120. [CrossRef]
6. Alshamrani, O.S.; Galal, K.; Alkass, S. Integrated LCA-LEED sustainability assessment model for structure and envelope systems of school buildings. *Build. Environ.* **2014**, *80*, 61–70. [CrossRef]
7. ISO. *ISO 14040: Environmental Management—Life Cycle Assessment—Principles and Framework*; ISO: Geneva, Switzerland, 2006.
8. ISO. *ISO 14025: Environmental Labels and Declarations—Type III Environmental Declarations—Principles and Procedures*; ISO: Geneva, Switzerland, 2006.
9. ISO. *ISO 14044: Environmental Management—Life Cycle Assessment—Requirements and Guidelines*; ISO: Geneva, Switzerland, 2006.
10. Ministry of Environment, Korea Life Cycle Inventory Database. Korea Environmental Industry & Technology Institute. Available online: http://www.edp.or.kr/lci/lci_db.asp (accessed on 3 December 2020).
11. Ministry of Environment, Korea Environmental Industry & Technology Institute. *Development of Integrated Evaluation Technology on Product Value for Dissemination of Environmentally Preferable Products*; Korea Ministry of Environment: Sejong, Korea, 2009.
12. Bang, J.; Tae, S.; Kim, T.; Roh, S. A Study on Developing BIM template based on Public Procurement Service Standard Construction Code to Improve the Efficiency in Carbon Dioxide Assessment of Buildings. *Archit. Inst. Korea* **2013**, *29*, 69–76.
13. Roh, S.; Tae, S.; Kim, R. Analysis of Embodied Environmental Impacts of Korean Apartment Buildings Considering Major Building Materials. *Sustainability* **2018**, *10*, 1693. [CrossRef]
14. Roh, S.; Tae, S.; Shin, S.; Woo, J. Development of an optimum design program (SUSB-OPTIMUM) for the life cycle CO2 assessment of an apartment house in Korea. *Build Environ.* **2014**, *73*, 40–54. [CrossRef]
15. Lee, S.; Tae, S.; Kim, T.; Roh, S. Development of Green Template for Building Life Cycle Assessment using BIM. *J. Korea Spat. Inf. Soc.* **2015**, *23*, 1–8.
16. Roh, S.; Tae, S.; Suk, S.; Ford, G.; Shin, S. Development of a building life cycle carbon emissions assessment program (BEGAS 2.0) for Korea's green building index certification system. *Renew. Sustain. Energy Rev.* **2016**, *53*, 954–965. [CrossRef]
17. Heinonen, J.; Säynäjoki, A.; Junnonen, J.; Pöyry, A.; Junnila, S. Pre-use phase LCA of a multi-story residential building: Can greenhouse gas emissions be used as a more general environmental performance indicator? *Build. Environ.* **2016**, *95*, 116–125. [CrossRef]

18. Choi, J. A study on the Building Energy Evaluation Using Land Plus Mass Parameter Family of Building Information Modeling at the Early Design Stage. *J. Archit. Inst. Korea* **2013**, *29*, 273–280.
19. Fan, S.; Skibniewski, M.; Hung, T. Effects of building information modeling during construction. *J. Appl. Sci. Eng.* **2014**, *17*, 156–166.
20. Laprise, M.; Lufkin, S.; Rey, E. An indicator system for the assessment of sustainability integrated into the project dynamics of regeneration of disused urban areas. *Build. Environ.* **2015**, *86*, 29–38.
21. Kang, L.; Kwak, J. Integrated Code Classification System for work Sections in Standard Method of Measurement and Construction Standard Specification. *Korea Inst. Constr. Eng. Manag.* **2001**, *4*, 81–91.
22. Hunkeler, D. Societal LCA methodology and case study. *Int. J. Life Cycle Assess* **2006**, *11*, 371–382. [CrossRef]
23. Lee, S.; Tae, S. Development of a Decision Support Model Based on Machine Learning for Applying Greenhouse Gas Reduction Technology. *Sustainability* **2020**, *12*, 3582. [CrossRef]
24. Roh, S. Life Cycle Sustainability Assessment Model of Buildings using Probabilistic Analysis Method. Ph.D. Thesis, Hanyang University, Seoul, Korea, 2017.
25. Lee, S.; Tae, S. The Greenhouse Gas Evaluation Methodology of Energy End Use based on Big-Data. *Int. J. Sustain. Build. Urban Dev.* **2020**, 78–93. [CrossRef]
26. Chae, M.S. Selection Model for Optimal Target Building for Green Remodeling Using Data-Mining Techniques: Focused on Multi-Family Housing Complex. Master's Thesis, Yonsei University, Seoul, Korea, 2017.
27. Ministry of Trade, Industry and Energy. *Energy Consumption Survey*; Ministry of Trade, Industry and Energy: Sejong, Korea, 2017.
28. Nakano, R.; Zusman, E.; Nugroho, S.; Kaswanto, R.; Arifin, N.; Munandar, A.; Arifin, H.S.; Muchtar, M.; Gomi, K.; Fujita, T. Determinants of energy savings in Indonesia: The case of LED lighting in Bogor. *Sustain. Cities Soc.* **2018**, *42*, 184–193. [CrossRef]
29. ISO. *ISO 21931-1: Sustainability in Building Construction—Framework for Methods of Assessment of the Environmental Performance of Construction Works—Part 1: Buildings*; ISO: Geneva, Switzerland, 2010.
30. Korea Energy Agency. *Guidance for Clean Development Mechanism*; Korea Energy Agency: Ulsan, Korea, 2018.
31. Peng, J.; Lu, L.; Yang, H. Review on life cycle assessment of energy payback and greenhouse gas emission of solar photovoltaic systems. *Renew. Sustain. Energy Rev.* **2013**, *19*, 255–274. [CrossRef]
32. Korea Environment Preservation Association. *Handbook for Environmental Labelling Certification*; KEPA: Seoul, Korea, 2013.
33. Singh, D.; Basu, C.; Meinhardt-Wollweber, M.; Roth, B.W. LEDs for energy efficient greenhouse lighting. *Renew. Sustain. Energy Rev.* **2015**, *49*, 139–147. [CrossRef]
34. Ministry of Environment. *Guidance for Greenhouse Gas Energy Target Management System*; Ministry of Environment: Sejong, Korea, 2017.
35. Lacirignola, M.; Meany, B.H.; Padey, P.; Blanc, I. A simplified model for the estimation of life-cycle greenhouse gas emissions of enhanced geothermal systems. *Geotherm. Energy* **2014**, *2*, 1–19. [CrossRef]

Article

BIM-Based Life Cycle Assessment of Buildings—An Investigation of Industry Practice and Needs

Regitze Kjær Zimmermann *, Simone Bruhn and Harpa Birgisdóttir

Department of the Built Environment, Aalborg University, A.C. Meyers Vænge 15, 2450 Copenhagen, Denmark; simoneb@build.aau.dk (S.B.); hbi@build.aau.dk (H.B.)
* Correspondence: rkz@build.aau.dk

Abstract: The climate debate necessitates reducing greenhouse gas emissions from buildings. A common and standardized method of assessing this is life cycles assessment (LCA); however, time and costs are a barrier. Large efficiency potentials are associated with using data from building information models (BIM) for the LCA, but development is still at an early stage. This study investigates the industry practice and needs for BIM–LCA, and if these are met through a prototype for the Danish context, using IFC and a 3D view. Eight qualitative in-depth interviews were conducted with medium and large architect, engineering, and contractor companies, covering a large part of the Danish AEC industry. The companies used a quantity take-off approach, and a few were developing plug-in approaches. Challenges included the lack of quality in the models, thus most companies supplemented model data with other data sources. Features they found valuable for BIM–LCA included visual interface, transparency of data, automation, design evaluation, and flexibility. The 3D view of the prototype met some of the needs, however, there were mixed responses on the use of IFC, due to different workflow needs in the companies. Future BIM–LCA development should include considerations on the lack of quality in models and should support different workflows.

Keywords: life cycle assessment (LCA); building information modeling (BIM); environmental impact assessment; sustainability; building life cycle; integrated design process; digitalization; greenhouse gas emissions; IFC; visualization

Citation: Zimmermann, R.K.; Bruhn, S.; Birgisdóttir, H. BIM-Based Life Cycle Assessment of Buildings—An Investigation of Industry Practice and Needs. *Sustainability* **2021**, *13*, 5455. https://doi.org/10.3390/su13105455

Academic Editor: Antonio Garcia-Martinez

Received: 28 March 2021
Accepted: 10 May 2021
Published: 13 May 2021

1. Introduction

The climate crisis necessitates an intensive investigation into reducing greenhouse gas (GHG) emissions. Here, buildings have a large reduction potential, as they are responsible for 38% of the energy and process-related GHG-emissions, globally [1]. To reduce the environmental impacts, life cycle assessment (LCA) of buildings is increasingly used. LCA is a widely used and accepted method of assessing the environmental performance of buildings. Moreover, LCA will in the near future become a mandatory requirement in several European countries such as Denmark, Finland, France, and Sweden [2,3]. However, the complexity and the time-consuming work related to LCA has often been considered a barrier [4–6], which now has to be overcome. Consequently, the efficiency potentials from using building information modeling (BIM) has gained attention in the literature [4,7], where several strategies for the workflow exist [7,8]. However, BIM–LCA is still at an early stage [7] and research on the topic is limited [4]. Some areas where research is lacking concern user-friendly platforms to assist in integration [4]. Further, to enhance interoperability between tools, integration methods with open file formats such as industry foundation classes (IFC) should be considered [4], which is currently less common in literature case studies [7].

The life cycle perspective is important because it includes considerations of material impacts. Due to previous years' political focus on reducing the operational energy use of new buildings, the impacts from materials have shown increasing importance [9–14]. The LCA method is described in ISO standard 14,040 and 14,044 [15,16] and, specifically for buildings, in the European standard EN 15,978 [17] from CEN TC350. Several nations have made their own method specifications considering the national contexts [2]. Life cycle assessment is used in sustainable certification systems [3,18,19], but several countries are considering or have decided to include limit values for GHG-emission in legislation, such as Denmark recently adopted [20]. Time and cost are part of the considerations from clients and legislators. Since some find the complexity of LCA high [21–23], this can be a barrier in the prioritization of LCA in the building industry. Especially for the early design stages, it can be an advantage to make LCAs quickly and often in order to support an iterative design process [24–26]. BIM can simplify the establishment of the life cycle inventory (LCI) for the LCA by eliminating the need to reenter information that is already available in the building model. Several studies have focused on BIM–LCA, but not through an industry perspective, where information is relevant for practical implementation in industry. The use of BIM in the industry is in continuous development. The use of BIM for public procurement is supported through EU directive 2014/24/EU [27], with national legislations [28]. In several countries, including Denmark, the use of BIM is required for public procurement of buildings, and the delivery must happen through IFC, which is an open interoperability standard [29,30] for architecture, engineering and construction (AEC), and facility management (FM). Several BIM–LCA studies have focused on IFC to support interoperability [31–34], however there is still a challenge with the poor design of the models [4,35]. This challenge could be addressed through a transparent and visual BIM–LCA approach. Here, some studies on BIM–LCA have focused on visualizing data from LCA directly in the model [36–38]. Further, some existing tools work with both IFC and visual interface such as the EveBIM in connection with Elodie [39], and the 6D-BIM-Terminal [34]. They use different approaches and focus on national contexts and specific situations, such as on the tendering stage. IFC and 3D view are also used in a Danish context, where a prototype has been developed to represent the workflow.

While prior studies have focused mostly on published academic case studies [4,7], BIM–LCA has become more common in industry practice. However, few studies on the practical use of BIM–LCA in the industry exist. The aim of this paper is to investigate this research gap by examining industry practices and needs in BIM–LCA. This includes the specific challenges related to the design of the models, and feedback on a prototype developed for the Danish context focused on the use of open neutral file formats and 3D view. "BIM" can be used to refer to more information-heavy tools and processes, but will in this study also include more simple, geometry modeling tools. Research questions in this paper are: (1) What workflow and challenges are related to BIM–LCA in industry practice? (2) What are needs for BIM–LCA in industry and are they met through the Danish prototype using open neutral file-formats and 3D view?

2. Background

2.1. Data Requirements for LCA of Buildings

While digital building models have an obvious advantage in creating the bill of quantities (BoQ), it is not the only data input required for an LCA. Following the terminology and method from European standard EN 15,978 [17], examples of additional data are operational energy and water use, service lives of products, transport, and maintenance and repair. These cover the different life cycle stages in order to determine the LCI, see Figure 1. Cavalliere et al. [40] have made an in-depth structure of relevant information to a BIM–LCA workflow. Furthermore, life cycle impact assessment (LCIA) has to be made, or an LCIA database for, e.g., building products, can be used. Different databases are available, and their use is typically connected to the choice of LCA-tool [41]. Since local adjustments in methodology for the building LCA exists [11], different data may be necessary depending on the context and goal of the LCA. These additional data can either be contained in the building model, or need to be added later on, for instance in an LCA-tool.

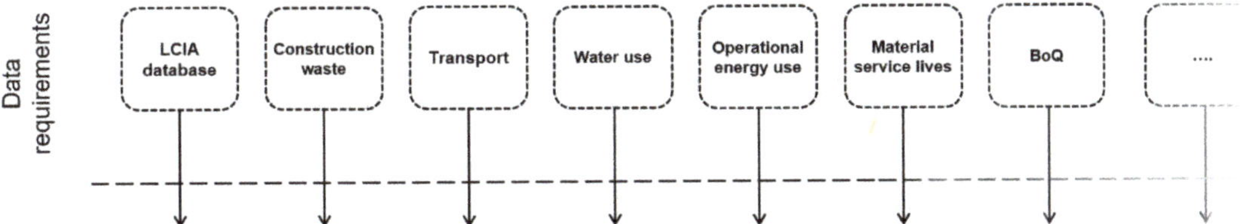

Figure 1. Examples of data requirement for LCA of buildings. Requirements and data structure can depend on the goal and context of the LCA.

2.2. Approaches for Integrating BIM and LCA

The literature distinguishes between adding environmental data into the building model, and only extracting information, such as the BoQ from the model [42,43]. Further, Wastiels et al. [8] categorize BIM–LCA integration into five approaches. These approaches also include the approach where LCA information is added to the model. This "enriched BIM" approach has the advantage that less information for the LCA needs to be manually attributed later on, thus supporting an automatic or semi-automatic workflow, which will greatly reduce human error [33]. Furthermore, centralizing data in the model can be an advantage in future uses of the model, such as facility management where an LCA may need to be redone [33]. Challenges for this approach are that the working environment for exchange of this information has to be established [33], including what information and where it should be attributed in the model, as well as how the data can be exchanged. Moreover, the work associated with changing a material in the model, in order to investigate different design solutions, may be larger than in an LCA software [8]. The most common approach in the literature is the "quantity take-off" approach [7]. Here, the BoQ is exported from the building model and then connected to an LCA software. The processes within the quantity take-off approach can range between manual and automated, depending on the use of different software for automation of the process. However, the manual process is the most common approach [7]. The nature of the approach is simple, but an iterative design process can be difficult, due to the manual processes involved. Further, the workload from manual processes can be extensive. The third approach from Wastiels et al. [8] is the "import of geometry into the LCA software", for example by using IFC for data exchange. An advantage of this approach is that IDs for the objects are used in the data exchange. This makes it easier to update the LCA without matching geometry and environmental data all over again. The fourth approach applies an intermediate "viewer" in a 3D environment, where information from, e.g., the IFC, is matched with environmental data. This approach has the same advantage with the use of IDs as the previous approach. Further, the match can happen within a 3D environment. For the previously mentioned approaches, there was no visual connection to the 3D environment of the building for the processes of matching data or visualizing results. The last approach also uses the 3D environment. This is the "LCA plugin" for the BIM software. Here, the BIM software automatically provides the 3D environment for matching and visualizing results dynamically for an iterative design process. The five approaches can be seen in Figure 2.

2.3. Data Exchange in BIM–LCA

The above-mentioned approaches are distinguished by their overall workflow; however, a crucial dimension is the type of data exchange. The data exchange within the tools available to the practitioners can limit their options for workflow.

Interoperability is typically the goal within data management between software solutions, to allow for easy exchange of data between software. Laakso and Kiviniemi [30] distinguish between the direct interoperability and open interoperability standards. An example of the open interoperability standard is IFC. The IFC schema is a standard, open, and vendor-neutral data model, describing the built environment [44]. Using a standard structure requires all relevant software to translate their data into the standard structure, thus creating a common language for all software to exchange data. For BIM–LCA, it is important to consider if the standard structure can contain the data you want to extract from your model, as described in Section 2.1. Using a standard data structure will always restrict how data can be described, and thus used in the building performance tools [45]. However, data interoperability using an open standard data structure has obvious advantages as it reduces the number of times data need to be translated [30], see Figure 3. In principle, the standard data structure can be used in all five approaches mentioned in Section 2.2, except the plugin solution.

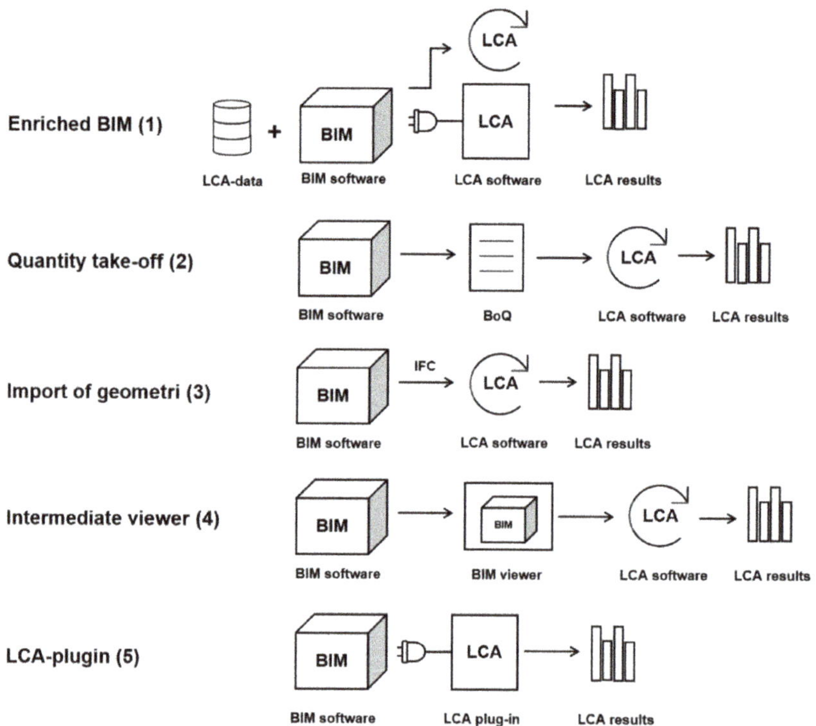

Figure 2. Five approaches to integration of BIM–LCA, as defined by Wastiels et al. [8].

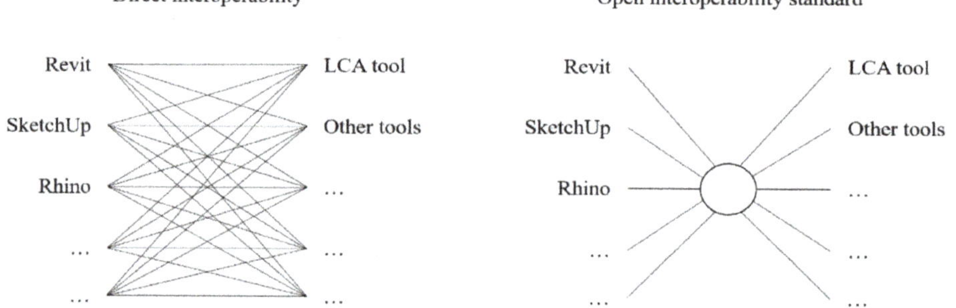

Figure 3. Data exchange from digital building model to LCA-tool using open interoperability standard and direct interoperability.

Alternatively, data can be transferred via direct interoperability, which requires some openness from the software providers in data structure [30]. This can be a challenge when proprietary data schemas are used. However, with an open data structure, data can be exchanged using, for instance, a file format to the target schema needed in the LCA. Open file formats that have been used for LCA are, for instance, xlsx [7]. These formats are typically used in approach 2 (quantity take-off), but can in theory be used for all above defined approaches, depending on the chosen data structure. The difference between this and the open standard data transfer is that it is not standardized, thus all transfers between tools, in principle, need to be made individually from each building model software to the LCA-tool, instead of using a common structure, see Figure 3.

Furthermore, software can provide the possibility of using plugins via an application programming interface (API) to exchange information with the software. An advantage of plugins is that it can add functionality to the original software, for instance, by visualizing results and receiving dynamic feedback on design changes within the building model environment. The plugin middleware can also select the specific data needed from the model for the data exchange with the LCA tool. Popular plugin solutions in the building sector are visual programming languages (VPL) [46], such as Grasshopper [47] and Dynamo [48], which make programming more available to architects and engineers. Plugin solutions can work alone without external dependencies, or as a bridge to an external LCA-tool. Approach 5 from Section 2.3 is defined as the plugin solution, however, a plugin can also work in connection with intermediate data schemas or formats. For example, VPL can be used to extract quantities and create an xlsx file, which can then be transformed to the LCA-tool data schema.

Some disadvantages of direct transfer are handling of software versions and errors in translation [30]. Furthermore, the plugin will only work with the specific software for which it is developed.

2.4. BIM–LCA at Different Design Stages

Data exchange in BIM–LCA can happen at different design stages where information in the models varies. Even within the same model, the level of development (LOD) can vary [49]. In early stages, the data for LCA from the building model is limited, and may not contain information on materials, for example. Conducting BIM–LCA at different LOD has previously been addressed in the literature [37,49–52]. Cavalliere et al. [49] and Röck et al. [37] suggest the use of predefined components based on the LCIA database for building materials when specific quantities are not known. For even earlier stages, average data for components or elements is suggested [49]. Predefined elements and components have also been suggested for early design LCA in general [2,21,24].

2.5. Prototype with Workflow for BIM–LCA

2.5.1. Context

A prototype has been developed in a Danish context as a possible workflow for BIM–LCA. The prototype only has some key features implemented, as well as some of the interface in order to give an idea of the functionalities. For the Danish Voluntary Sustainability Class [53] and the Danish adaption of DGNB (Deutsche Gesellschaft fur Nachhaltiges Bauen) [54], it is mandatory to use the environmental product declaration (EPD) or use the LCIA database, Ökobaudat [55]. Thus, one of the main goals for the BIM–LCA integration process is to gain information on material quantities and match the information with environmental data. The information is connected to the Danish national tool, LCAbyg [2,56].

2.5.2. Workflow

A prototype for BIM–LCA was developed to meet some of the challenges associated with poor design of models. The prototype was developed using the "viewer" approach as described in Section 2.2., but also closely related to the "import of geometry" approach, because the prototype is closely connected to the dedicated LCA software. The general idea of the developed prototype is: (1) the use of standard and open file-based exchange with flexibility in data input to support use across different design stages; (2) create a visual interface in order to enhance the quality and documentation of BIM-based LCA, and to support an iterative design process. The workflow is shown in Figure 4. From the building-model software, the data is exported to an open file format. This format is imported into the prototype, where the necessary information is added in order to perform the LCA, including matching the BoQ with LCIA data. The matching of BoQ with LCIA data happens manually or semi-automatically in the 3D environment based on the information available from the model, and the library of LCIA data. The semi-automatic process consists of suggestions of matches based on previous matches or material names. Further, objects with identical material composition can be grouped together and matched to LCIA data using names, classification, IFC-structure, shape, etc. This process can be further automated if information from the LCIA library elements have been implemented in the building model, following the approach of "enriched BIM". In the 3D view, the object placement and quantities can be visualized. The LCA is carried out in the Danish LCAbyg-tool. LCAbyg is connected to the prototype through direct interoperability in python, using JSON-format to exchange information with LCAbyg. The prototype can be used to visualize results from the LCA directly in the 3D-model.

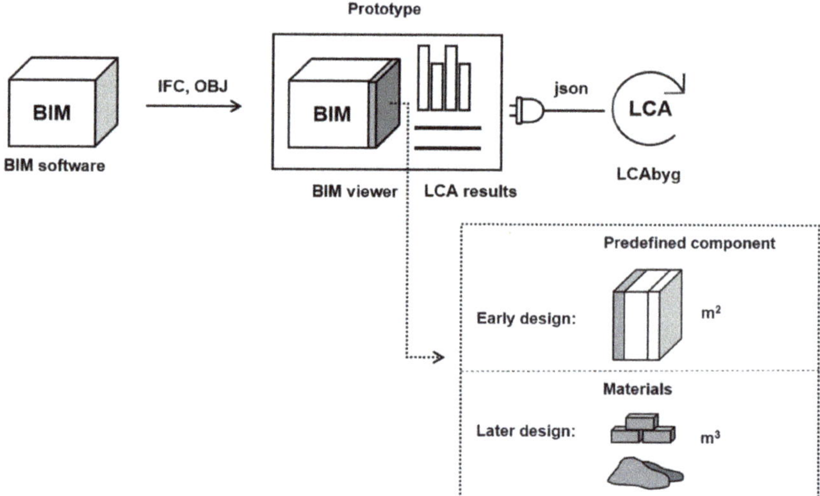

Figure 4. Workflow for BIM–LCA in the prototype. At different design stages, it is possible to work with different types of available information from the model.

2.5.3. Use across Different Design Stages

Due to variations in LOD of models during a building project, the prototype uses predefined components as described in Section 2.4. The user can match predefined components with the quantities in the model. All quantities are calculated and available in the prototype tool, thus it is possible to use the quantities that are relevant at the current design stage of the building model. In earlier stages with low LOD, the material information is likely not modelled. Here, environmental data for predefined components or elements can be matched with areas extracted from the building model. Predefined components are a part of the library in the Danish LCAbyg tool [57,58]. At later stages, the specific material quantities can be extracted from the digital model, or added within the prototype. This is illustrated in Figure 4. Results are provided through the LCAbyg tool.

2.5.4. Open File-Based Data Exchange

Open, file-based exchange was chosen as the data exchange in order to support a wide range of software for the digital models without creating middleware for each individual building model software.

For the prototype, two file formats have been selected for the data exchange: the IFC schema for the more complete data exchange, or OBJ for a limited data exchange. IFC is an open standard data model for AEC and FM, and can be represented through a file-based exchange [30]. OBJ is an open file format for describing 3D geometry. OBJ is strictly geometry, whereas IFC contains object based information which can store a large variety of data on the building. The information available in the IFC depends on the Model View Definition (MVD) [44] and can be different depending on the used building model tool, or the selections the user makes when they export their model to IFC. A specific MVD can be made for the data exchange, and has been developed in other studies [32,33,59]. However, for now the prototype will not require any specific information in the IFC. This way, the tool will be able to support all IFC models, no matter how they have been processed previously by software and users. The OBJ can act as a practical alternative to IFC because the process of import and export is faster than IFC, and the limited data exchange of OBJ will likely be enough for the early design stages where geometry is the only information available in the building model. Furthermore, export to IFC is not always accessible in the design tools (see Table 1). IFC and OBJ both use unique IDs for objects, making it possible to have an iterative process in the building design, without repeating the manual processes, as described in Section 2.2.

Table 1. Export options for Industry foundation classes (IFC) and geometry file format OBJ from different model software.

Model Software	IFC	OBJ
Revit	x	x
Rhinoceros	x [2]	x
Sketchup	x [1]	x [1]
ArchiCAD	x	x
AutoCAD	-	x [2]
Vectorworks	x	x

[1] Not available in the free version. [2] Requires purchasing of plugin.

2.5.5. Visual Interface

The visual interface in the prototype was achieved through an interactive 3D view of the building. See Figure 5. In this view, it is possible to navigate similarly to other 3D tools (zoom, rotate, etc.). When the user targets an object, the available information for the LCA is shown, such as quantities and material information. IFC and OBJ can both provide 3D-object information, necessary to visualize the building. The visual interface is where the BoQ is matched with LCIA data. Further, the 3D interface can be used to visualize results from the LCA. It is also meant to give a better understanding of the origin of the BoQ, and if there are collisions, missing or wrongly categorized objects, or other errors. The modelling errors become easier to find when they are visualized in the 3D model. The prototype calculates the quantities, but the user can also choose to use quantities from the original building-model software if they are included in the IFC. Moreover, it is always possible to overwrite the quantities or other information from the model.

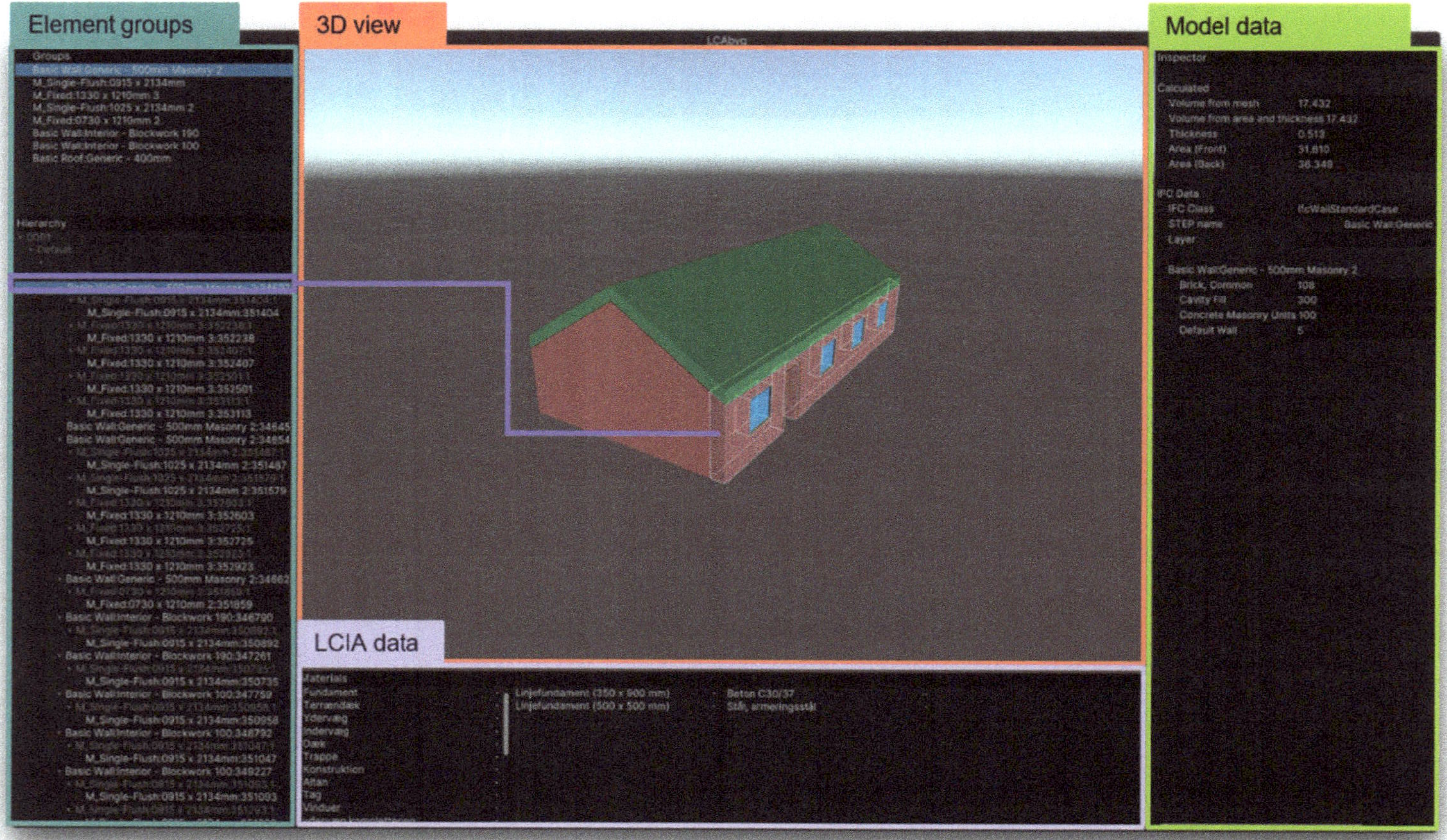

Figure 5. Prototype interface with 3D view of the building.

3. Materials and Methods

Qualitative Interviews

Data in this paper is based on qualitative in-depth interviews with companies who perform LCA of buildings. The goal of this method is to understand the company perspective on performing LCA of buildings, such as their current practices and motivation behind them, as well as demands for better workflow and feedback on the presented prototype.

The qualitative interviews consist of eight semi-structured interviews with companies in the Danish building sector, who offer LCA of buildings as a service. The companies were selected to represent a variation of company types with consultant services, from architect, engineering, and contractor firms. For further details, see Table 2. Contact with the companies had already been established through previous projects with LCA in the building industry. Prior to the interview, the themes of the interview were given to a contact person in the company and they were prompted to bring relevant informants from the company to the interview. Further, the questions were sent to the companies prior to the interviews, to give them an opportunity to prepare, or ask others in the company if they didn't have the answers themselves. The semi-structured interview focused on the following questions. Prior to the last question on the list, a presentation of the developed prototype was given:

- Which digital building model tools ("BIM") and LCA-tools do you use today?
- How is the BIM–LCA workflow in the company today, and why?
- How do you work with BIM in relation to LCA? E.g., use of discipline models;
- What challenges do you face in BIM–LCA?
- What is most important for a good BIM–LCA workflow? E.g., quick, automation, ease of use, transparency/quality assurance, flexible workflow, precision of data, visual/3D view, evaluation of design solutions, understand LCA and material impacts.
- Does the prototype satisfy these important aspects? What does it meet/doesn't meet, and why?

Table 2. Overview of the company type and informant profiles in the eight semi-structured interviews.

Interview	No. of Informants in Interview	Profiles	Company Type	No. of Employees in Denmark (in Ranges)
A	2	Engineers	Consulting engineers and architects	3000–3999
B	2	Engineer and design engineer	Consulting engineers and architects	1000–1999
C	3	Engineer and design engineer	Consulting engineers	100–199
D	1	Engineer	Consulting engineers	500–999
E	2	Engineer and architect	Consulting architects	100–199
F	1	Architect	Consulting architects	0–99
G	2	Engineer and architect	Consulting engineers and architects	3000–3999
H	2	Engineers	Contractor and consulting engineers	1000–1999

The interviews were analyzed and categorized using a combination of deductive and inductive coding technique [60]. The deductive coding technique is based on the theoretical background, and the inductive coding technique arose from informants discourse. The purpose of this is to understand the companies' workflow in relation to the existing literature on, e.g., the BIM–LCA approaches presented in Section 2.2, while including themes that arose from discussions with informants, such as the challenges they meet in BIM–LCA.

The eight interviews were comprised of 15 informants and were carried out in November and December 2020, and January 2021. The informants were engineers, architects, and design engineers. They covered informants with knowledge on LCA of buildings and, for some companies, informants that work across disciplines with a focus on sharing and using digital building information.

As stated above, the companies represent a broad variety of professional profiles and companies. The companies cover large and medium size companies, but not small or one-man businesses. Due to the size of the interviewed companies, they cover a large part of the Danish AEC industry, but it should be noted that the building industry in Denmark also consists of many smaller companies [61,62]. For this research, smaller companies were considered to have too little experience in BIM–LCA to give valuable input. The interviewed companies were chosen due to their knowledge and practical experience in performing LCA on buildings. The selected companies are part of an LCA expert group, who are consulted in relation to the development of the national tool for LCA on buildings, LCAbyg [2,56]. Due to their advanced knowledge in comparison to many other, and smaller, companies, their experience can inform in more detail on practical workflow and challenges as well as demands.

4. Results

4.1. BIM–LCA Workflow in Companies

The most commonly used BIM–LCA workflow in the companies is the quantity take-off approach, as presented in Section 2.2, and a few of the companies have started development on the LCA-plugin approach for the BIM software. However, the companies work differently within the approaches. Figure 6 illustrates how the individual companies work within the two approaches. All companies use direct interoperability for data transfer, but with some differences in approaches. Three of the companies use export of schemas from the BIM software, Revit, to Excel, in order to create the BoQs from the BIM. At times, company H creates the BoQs using a Dynamo script from Revit to an xlsx-file, along with company C and D. Here, company B uses a C# script for the same process of creating an xlsx file. All the mentioned companies manually transport the BoQs in the xlsx file into the LCA software, LCAbyg, where the LCA is done. However, company D typically uses their own excel tool for the LCA, and only does the final calculation in LCAbyg.

Company E uses a semi-automatic BIM–LCA workflow, where the BoQs are created from a Rhinoceros-model (Rhino) using a Grasshopper script. A library with predefined constructions can be linked by the user to the BoQs in Grasshopper, and JSON (JavaScript Object Notation) files are created according to the target schema in the LCA software, LCAbyg. Company A also uses a semi-automatic BIM–LCA workflow, where the BoQ in excel is created from a Revit model using Dynamo or export to Excel. In the Excel file, they can match BoQ with IDs for LCIA data. Based on the xlsx-file, a script transforms the data to xml files according to the target schema in the LCA software, LCAbyg.

Currently, some of the companies are developing the LCA-plugin approach for the models, to use in the early design stage. Company C is working on a solution for early design stage, using Rhino and Grasshopper, and company D is working on a tool using Revit, Power-BI, and matching via classification codes. These are still under development, and have not been included in Figure 6. Company G has recently developed a plug-in solution for the BIM software, Revit, where LCA results can be shown dynamically as the user edits the Revit model. The environmental impacts from a library with predefined constructions are linked to the keynotes in the Revit model.

4.2. Data Used for BIM–LCA

A BIM model is naturally used in the BIM–LCA workflow; however, several other data inputs are used within the companies. Figure 7 illustrates the different sources of data used for building models and how, in most cases, this information is supplemented with additional data.

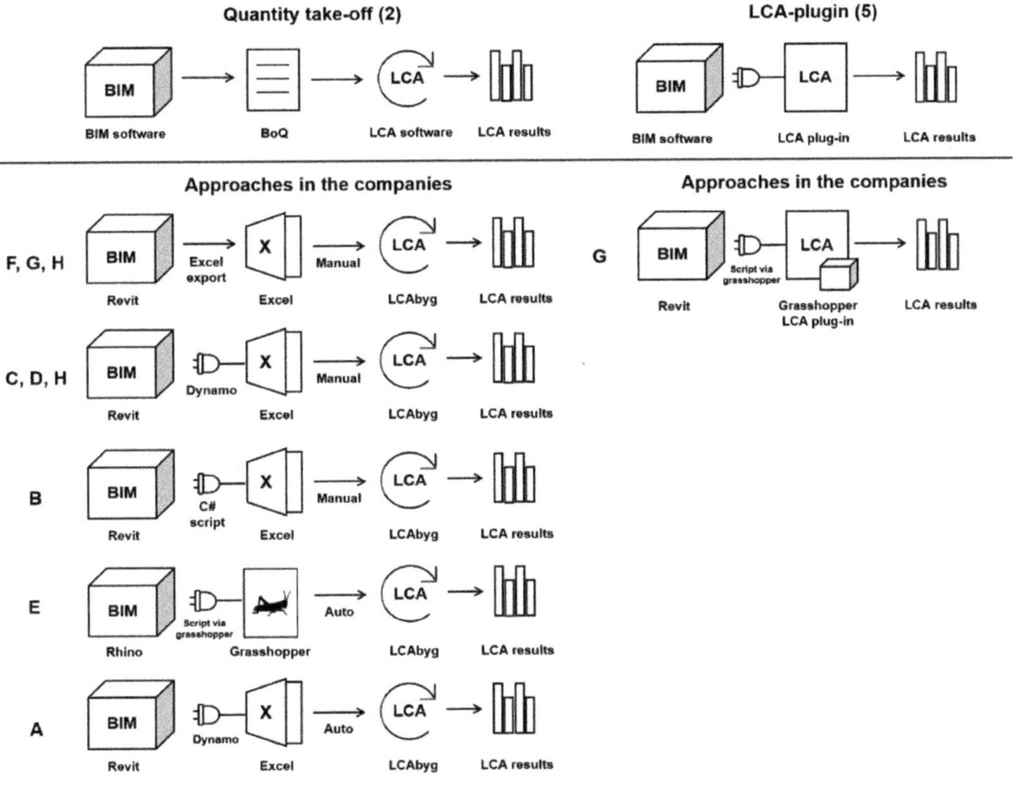

Figure 6. Detailing of BIM–LCA approaches used in the companies.

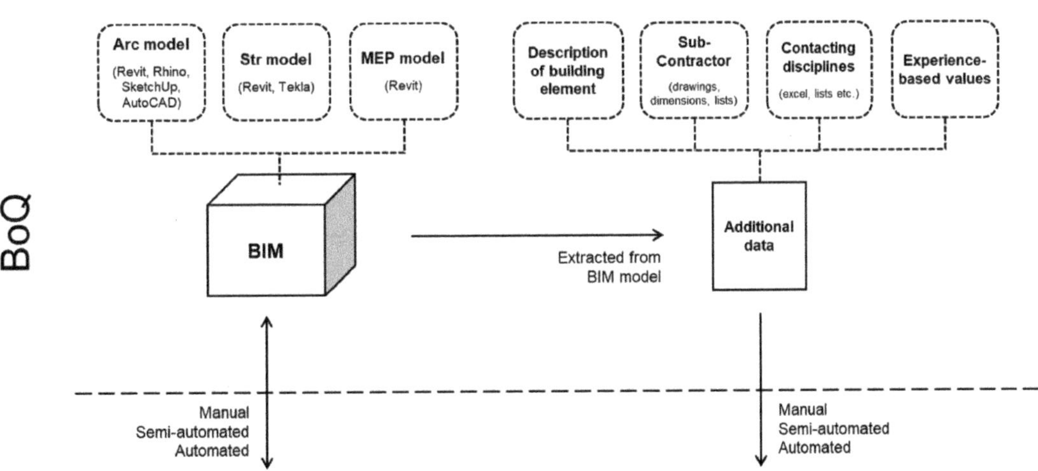

Figure 7. Data sources used in the companies.

Different models exist during a project, and this is reflected in their use within the companies. In general, all the companies mention Rhino as a tool that is used in early design stages, where Sketchup and AutoCAD is also mentioned in a couple of the companies. The Rhino models are in some cases used for the LCA as illustrated in Figure 6. In the more detailed stages, all companies use Revit. They describe Revit as almost an industry standard when modelling in the project design stage. The companies work with different discipline-oriented models in Revit: an architectural model, structural model, and mechanical, electrical, and plumbing (MEP) models. All companies use the architect model for the LCA, but only two companies mention that they extract the data from the structural model and the MEP models to perform the LCA, and only in the detailed design stage.

To supplement data, and to fill the data-gap from only using the architectural model, the companies mentioned additional data sources. These include descriptions of building elements, data from sub-contractors, and gathering data from the discipline groups such as the structural or HVAC (heating, ventilation, and air conditioning) engineer. Two companies also mentioned the use of experience-based values from earlier projects or the literature to supplement in earlier stages, when data is not available. The use of descriptions of building elements is mentioned by company B, C, and H for LCA in the early design stage, when information in the model is limited or when it is not defined in the BIM. Element details are gathered from the supplier, for example the concrete element supplier, because they have more detailed information on the elements. If information or data are missing in the BIM model, the companies contact the discipline groups to collect the missing information. An example of this is company A, who collects information by providing the different discipline groups with Excel sheets, where they can fill in the data.

4.3. Challenges in BIM–LCA

During the interviews, the individual companies were asked which challenges the company faces when making LCA from the building models. The challenges are listed in Table 3, where they are separated into eight overall challenges.

Table 3. Challenges of BIM–LCA mentioned by the companies.

Challenges	Comments
Lack of building-model management for a collaborative process	• Those who need information from the model (e.g., quantities of materials) are not the ones who model it (A, F, C); • Modelling starts very late in some projects, especially the structural model (G); • The consulting engineer may not design the ventilation system themself, but puts it out to tender. Thus, they don't have the model (G, F); • Contractual issues means that they cannot edit in, e.g., the architectural model (D); • No minimum demands for LOD on material information exists (A); • No common understanding or standard for extraction of quantities (F); • Challenging to motivate other actors to include materials in the Revit model, when it takes long, and gives no value to the one who does the modelling (F); • Lack of responsibility of the quantities in models (A);
Workflow errors	• Human error when manually typing into LCAtool from 8–10 different Revit schedules (F); • Extracting quantities from Revit is a black box, where it is not possible to see if anything is missing (F); • Difficult to check the models for errors, when someone else has made the model (A); • Paint areas are wrong, if the suspended ceiling is not accounted for (A);

Table 3. *Cont.*

Challenges	Comments
Lack of data availability and quality in models	• The data in the models are not good enough to form a basis for a good BIM–LCA integration (A); • Issues with extracting correct quantities from the models (F), specifically volumes (D); • The models are modeled incorrectly in terms of extracting quantities, although the graphical representation of the model looks correct (F); • Quantities will always be incorrect to some degree (C); • 10–15% of the model is not modelled correctly (G); • Quality of the modelled elements vary (G); • MEP model is not used for the LCA because it is not good enough. They collect the quantities on a list from the engineer (G); • Structural model from the consulting engineer is not as good as getting information from element supplier (G); • Not all materials are modeled in the model, e.g., steel in the plaster wall (B); • Detailing is not very high in the Revit model, e.g., they don't model reinforcement or holes in slaps (G); • Detailing varies (C); • Not all data are available in the model and likely never will be (C); • Often there is no structural or MEP building model (more often in office buildings, as they have higher demands) (G); • Information is not in the Revit model, only geometry (E); • Materials are not in the models (D, A);
Modeling errors	• Delta beams, piping, etc. are drawn as solids, resulting in the wrong volume (A); • No reinforcement in concrete elements (A); • Errors in model, e.g., internal walls are modelled as external walls (H, C) or as wall instead of foundation (A); • Some elements are modelled doubled, because several disciplines have modelled them (e.g., architectural and structural models both include structural elements). There is a risk of double counting (A, F, G); • Wrong dimensions of elements (A); • Columns drawn through slaps, giving the wrong volume (A); • Windows drawn as curtain walls (A);
Variations in the structure of models	• The structure in the models varies (B, D), and the model they get from the architect is structured differently each time (C); • The structure of the objects in the models varies (B), e.g., variation in the construction of the floor; with or without deck, etc. In the early design stage the objects are modelled as generic elements, while in the detailed design stage the building elements are modelled with all functional layers, e.g., ceiling, floor; • Modelling is different in other nations (G);
Data exchange and matching model-data with LCIA data	• Quantity outputs units from models are sometimes difficult to use for LCA, e.g., "pieces" of stairs (G); • Matching quantities with LCIA data from LCAbyg (C); • It is a challenge to create generic plugin scripts for all models as they are modeled differently. They always need to adjust the VPL/script (D); • Difficult to predict the future and thereby develop tools or a workflow for future processes (A); • Oversimplified or too user-friendly tools (F); • Issues with stability and/or workflow of different VPL (A, B, C, F, H);
Manual workflow and large models	• Time consuming with manual BIM–LCA workflow (F, G); • Extracting quantities/checking data is the most time-consuming process (D, A); • The large number of elements in a model makes it a time-consuming process (A); • Too much information in the models can make them slow to work with (D).

Some of the most commonly mentioned challenges are the lack of data availability and quality in the models used to establish the BoQ. An architect mentions that the models have not been made for the purpose of quantity extraction, but with other aspects in mind, thus the quantity take-off is wrong. It is also mentioned that some of the discipline models, such as structural and MEP, often do not exist, or are not reliable for quantity take-off. Further, the detailing varies, but some materials are simply not included in the model, such as reinforcement, and steel in plaster walls. Several mentioned that it is not likely that quantities will ever be completely correct in the model. Model errors are listed as a separate challenge in Table 3, however, they only contribute to the lack of quality in the models.

Furthermore, the structure and classification of the models can vary a lot, which can influence the data exchange. For instance, if a plugin expects a certain structure, but the model doesn't have this structure. When matching the BoQ to LCIA data, a common challenge mentioned is matching the units, as they may not align. It is also a source of human error, if the match is done manually. Some mention that the manual processes are time-consuming. This also includes manually checking the quantity take-off, due to the above-mentioned lack of quality.

To some degree, these challenges are a result of the lack of management or standardization of the models in relation to LCA, where some mention the lack of method for extraction of quantities, requirements for input of material information, and good-quality models at the time that they need them for the LCA. Further, those who make the LCA are often not the ones who make the building models. Therefore there is a lack of incentive for modeling for quantity take-off, or a lack of responsibility of the quantities in the model which is needed in this collaborative modelling work.

4.4. User-Perspective on Integration and Response to Prototype

The informants were asked about features for the integration process that they found important, and afterward they were presented with the prototype from Section 2.5 and provided feedback. Both of these results are shown in Table 4. In terms of important features for the BIM–LCA, one of the informants said that the integration should help solve the data issues from BIM. This refers back to the challenges, mentioned in Section 4.3, where several companies questioned the quality of their models, and their completeness. The 3D view was mentioned as a positive feature in connection to transparency of data from the model. Due to the quality of the models, they need to check the quantities, thus the 3D view will help them understand the origin and calculation of quantities, and to see if there are collisions of elements. The 3D view was also mentioned in relation to visualization, where several companies suggested it and found it to be a positive feature in the prototype. In general, six out of the eight companies mentioned the positive in a visual interface for the BIM–LCA integration. They mention its positive effects on communication and discussing results with different actors of the projects, especially at early design stages. Two engineering companies stated that they do not necessarily need a 3D view, as they were worried that the integration process would take longer. In terms of ease-of-use, some worried that the general workflow in larger models might be complex, if they need to review and match all this data with LCIA-data. However, some said that the grouping and filtering of elements can be used to manage the data.

Automation was another theme several of the companies found important. One of the informants mentioned that the models will likely always be wrong, but they still see potential in automating 80–90% of the process. Another informant mentions that automation is valuable, because humans make mistakes, and human mistakes are much harder to find. Automation also has relevance in terms of efficiency, where they currently spend many hours extracting quantities and go through several steps to make the LCA. To make automation easier, one informant suggests to "enrich" the BIM with information that can automatically match to the LCIA data. When presented with the prototype, one found it positive that the IDs from the IFC would make it easy to update the model, while another mentioned the lack of dynamic or parametric features.

Table 4. Important aspects of the BIM–LCA integration process mentioned by the companies, and their comments on the prototype.

Important Properties for Integration Process		Comments on Prototype
Ease of use (G, H)	• Everyone should be able to use it. It should be simple (G); • Help solve the issues in data from BIM (H);	Cons: • In a building model, they have 300 different Revit "families". This might be too much work/too complex to work with in the prototype. (A, B, F); • Worried that the tool cannot handle larger models (that the program might crash) (D, G);
Visual interface (A, B, C, D, E, H)	• Important for early design stages (D, E); • Interface with 3D-model (A, E, H,); • To communicate and discuss results of LCA with other actors (B, C);	Pros • 3D interface (C, D, E, F, H); • Communicate result to client (B); • That you see a 3D view of the actual building, you are working on, not just a generic model. (F); • You can see the objects you have matched to LCIA data vs. those you haven't yet (F); Cons: • It might be faster to manage the data without the 3D view. They don't always need a 3D view, if it takes more time (B, C);
Evaluation of design solutions (B, C, G, E, H)	• Show where to focus the optimization, e.g., the largest impacts (H); • Comparison of building elements and materials (B); • Comparison with their own or certification references/benchmarks for buildings (B, H); • Important for early design stages (E);	Pros: • Comparison of design solutions (B);
Transparency of data from the building model (A, B, C, H, F)	• They need to assess the quality of the model, therefore, they need to see how BoQ is connected to the information from the building model (H); • The models will likely always be wrong, so they have to check it (A, B); • Possible to see where there are changes or new objects, when you update the model (C); • Highlight obvious errors, e.g., the building being much heavier than similar building. (F); • 3D visualization with names and thickness of elements (H);	Pros: • Quality assurance of data, especially when elements can be filtered/grouped together (G); • See all the building elements in 3D view (H); • Easy to understand the origin of quantities with 3D view (D, A); • You can see how areas are calculated due to the 3D view (C); • Quantities are also calculated within the tool, not just quantities from Revit (F); • You can more easily see if you are missing element/materials (F); • Collision control (F); Cons: • Too complex in larger models to do quality control (B);
Precision and completeness of BoQ data (B, D, E, F)	• The LCA should have large detailing already at early stages. Therefore you should be informed of missing elements, e.g., ventilations systems (F); • Quantities from the building model should be correct (D); • Important at later stages (D, E);	

Table 4. *Cont.*

Important Properties for Integration Process		Comments on Prototype
Quick/automation (B, C, D, H, F)	• Currently, there are too many steps before the final LCA can be made (H); • They spend many hours extracting quantities (B); • Retrieve quantities from the model and update them automatically when the model changes (B); • The matching of BoQ with LCIA should be remembered when the model is updated (B); • If 80–90% of the process in the future will be automated, it will be a great help (B); • To make an automatic match of quantities with LCIA data, LCIA should be included in Revit/IFC (B); • Automation of the processes is a good idea, because human errors are difficult to find (F); • Important for early design stages (D);	Pros: • Easy to update the model, due to ID's when using IFC (F); • The prototype tool contains the library used in the Danish tool, LCAbyg (H); Cons: • Not dynamic or parametric (E); • If the architect deletes a wall and draws a new wall, it will have a new ID, and then you cannot as easily update the LCA anymore (C);
Flexible workflow in terms of data sources (A, C, E, F, H)	• Import of IFC and Revit, as this is what is most commonly used in the industry (H); • Not certain that Revit is what we use in the future, therefore more file formats should be possible to use (F);	Pros: • Can possibly solve the issue with the uses of different building model tools in the industry (H) • Neutral file format (H); • The possibility to use areas as quantities and match with LCIA-data for predefined elements, as an alternative to specific quantities such as m3, kg. (D); • Choose what data, they use from the models, because they know that some information is not correct (A); • Possibility to overwrite and adjust quantities and structure from the building model in the LCA (G); Cons: • They prefer that it is made specifically for Revit, because they mainly use Revit (D); • They might prefer exchange via files such as 3DM or MWD as it might be faster than IFC (C);

Five companies also find the flexibility of data sources important. One mentions that IFC and Revit are the most commonly used data sources in the industry, and thus should be supported in a tool for BIM–LCA. Another mentioned that it is not certain that Revit will be the main tool in the future, therefore other data sources should be supported. When presented with the prototype, some found the use of a neutral file format positive, while others preferred to focus on Revit or use different file formats than IFC and OBJ. Some had a general experience of "loosing" their data when they had previously used IFC in their work. In the prototype, some found the flexibility positive; in terms of choosing only the data that they find relevant from the model, as well as the type of quantities relevant to the stage of the project, e.g., choosing areas instead of kg and m3 for early design.

Evaluation of design solutions was also important to consider in BIM–LCA for several of the companies, in order to get instant feedback on design solutions and whether or not they meet certain benchmarks. Four of the companies also mentioned that precision of data is important, including completeness of data already in the early stages, such as by including installations. Referring back to the challenges in Section 4.3, this information may not be available in the model and thus have to be added in the BIM–LCA process.

5. Discussion

5.1. Data Management

The companies interviewed for this study only used the model to store data related to extracting the BoQ. However, storing more LCA-related data in the model can reduce human error, support automation, and facilitate better use of the models across the life cycle of the building [33]. Moreover, it complies with the concepts surrounding BIM, which focus on information sharing and collaboration across the building life cycle. However, the workflow for this "enriched BIM" first needs to be established [33] and may vary depending on the goal and context of the LCA, as well as the structure used in the model. Further, if the model includes environmental data, it can be a challenge to manage if it is up-to-date [63]. Inclusion of environmental information in the BIM and using the IFC-viewer workflow has been tested in the literature before, with more focus on the later stages [59]. However, the process is associated with practical challenges, because even though IFC can contain this information, some properties, attributes, and entities are not available in industry BIM [59,64]. Further, the IFC schema still needs to be improved to allow information for a full LCA [33].

Despite only using BoQ data from the model, the companies are met with challenges related to the quality of the model and many use supplementary sources to complete or detail the BoQ. Poor design of models for LCA and life cycle performance has been recognized in the previous literature [4,35], and is confirmed and specified in this study. While future legislation demands for LCA might improve the collaboration related to quantities in the models, several companies expressed that it is not realistic that the models become perfect in terms of quantity extraction. An issue therefore lies both in how the BoQ data from the models can be improved, and what expectations regarding the precision of BoQ is expected from the building LCA at different stages. Automation could be a possible solution to improve upon the data quality, such as automatically adding reinforcement in concrete elements. However, automatic or semi-automatic approaches can also be imprecise and reduce transparency in the process. In terms of the expected precision of the LCA, the practitioners will likely need clear guidance regarding this aspect in relation to benchmarking their building.

In early design stages other strategies can be used, such as matching quantities with predefined elements, as suggested in this article as well as in previous studies [2,21,24,37,49].

5.2. Tool for BIM–LCA

The prototype for the Danish context includes the visual interface in correlation with conducting the building LCA. The companies were generally positive towards the 3D view in the prototype for both transparency of data and visualization of results. Some of the companies were also working towards their own plugin approach with 3D view, especially for early design stages. In the development of the prototype, it could be relevant to be inspired by the plugin–workflow, for instance by allowing the user to modify the geometry in the prototype to achieve the same dynamic effects, and test different designs. A challenge in the plugin–solution is the dependency on one specific building model tool. The companies from this study mainly use Revit, and some therefore preferred a direct data-exchange for this software. However, for the early design stages, it is more common to use a variety of tools, and some companies also expressed the positive in using neutral file formats in order to support a variety of modelling tools. It is likely that some companies will want to optimize internal processes, and thus develop their own tools, while others will require ready-to-use software. Software providers and policy-makers should therefore allow for different workflows, and provide a clear description of method.

5.3. Limitations

While the interviews can give detailed insight into workflow, challenges, and demands for BIM–LCA in industry practice, it should be noted that this study is a qualitative study with a limited sample size. Thus the results from the study represent the experience in eight different companies in Denmark. The companies cover a large share of the Danish AEC industry due to the large size of some of the included companies. The companies are of varying size, however, small and one-man businesses are not represented in the interviews, because it was assumed that they would have limited experience in the subject. Omitting the small companies can potentially have an influence on the informant's feedback on the prototype. This is because small companies can be more dependent on ready-to-use tools, such as the prototype, because they have less resources to develop their own integration of BIM–LCA. The prototype facilitates an integration process where all models, independent from which software the model is created in and how it is structured, can be used for BIM–LCA. Future development of the prototype should therefore include considerations of smaller companies.

6. Conclusions

This paper has provided insight into industry practice of BIM–LCA through eight in-depth interviews with consulting and contracting firms. All the companies use a quantity take-off approach for the BIM–LCA and some have recently made, or are currently developing, plug-in solutions. Nevertheless, due to the lack of quality in the models, it is often necessary to supplement the model-data with data from other sources, such as element descriptions and contacting engineering disciplines and subcontractors. The lack of quality and variations in modeling are dominant challenges mentioned by the companies. Many of these issues points back to a management of the models, which is not optimal for quantity take-off. In the future, the quality of the models may improve due to legislations in, e.g., LCA, however, some degree of inaccuracy should always be expected, especially in early design stages. For the integration of BIM–LCA it should therefore be considered how the inaccuracy is dealt with. Moreover, to which degree automation can be incorporated in the process. For legislation and benchmarking, the level of detail expected for the LCA should be clearly defined.

The informants also provided needs for BIM–LCA and evaluated a prototype for BIM–LCA in a Danish context with the use of open neutral file formats and a 3D view. The companies considered several aspects important in BIM–LCA, including visual interface, transparency of data, automation, flexibility of data sources, and easy access to evaluation of design solutions. Many considered the 3D view in the prototype valuable for transparency and communication, but some questioned its efficiency and use for their larger models. The prototype uses open and neutral file formats such as IFC and OBJ for the data exchange, which garnered mixed responses from the companies. Some valued the flexibility it can provide in terms of using models from different software, while others preferred optimizing the direct data exchange to their predominantly used tool, Revit. Companies will have different resources and goals, and thus different needs in relation to workflow for BIM–LCA. Specifically smaller companies will likely benefit from ready-to-use solutions such as the prototype, because there are no requirements to the structure of the model, or the software used for modeling. A strategy for software developers and decision-makers can therefore be to allow for different workflows, but provide transparency of results and clear descriptions of method.

Author Contributions: Conceptualization, methodology, validation, R.K.Z.; formal analysis, R.K.Z.; investigation, data curation, R.K.Z. and S.B.; writing—original draft preparation, R.K.Z.; writing—review and editing, R.K.Z. and H.B.; visualization, S.B. and R.K.Z.; supervision, H.B.; project administration, H.B.; funding acquisition, R.K.Z. and H.B. All authors have read and agreed to the published version of the manuscript.

Funding: This research was funded by Realdania, grant number PRJ-2019-00308.

Institutional Review Board Statement: Not applicable.

Informed Consent Statement: Informed consent was obtained from all subjects involved in the study.

Data Availability Statement: Data sharing is not applicable to this article.

Acknowledgments: The authors would like to thank the participants in the interviews. For the software programming, the authors would like to thank Christian Zimmermann and Christian Grau Sørensen.

Conflicts of Interest: The authors declare no conflict of interest.

Glossary

AEC	Architecture: Engineering and Construction
API	Application Programming Interface
BIM	Building Information Modeling
BoQ	Bill of Quantities
DGNB	Deutsche Gesellschaft fur Nachhaltiges Bauen
EPD	Environmental Product Declaration
FM	Facility Management
GHG	Greenhouse Gas
HVAC	Heating, Ventilation, and Air Conditioning
IFC	Industry Foundation Class
JSON	JavaScript Object Notation
LCA	Life Cycle Assessment
LCI	Life Cycle Inventory
LCIA	Life Cycle Impact Assessment
LOD	Level of Development
MEP	Mechanical, electrical and plumbing
VPL	Visual Programming Language

References

1. United Nations Environment Programme. *2020 Global Status Report for Buildings and Construction: Towards a Zero-Emission, Efficient and Resilient Buildings and Construction Sector*; United Nations Environment Programme: Nairobi, Kenya, 2020.
2. Kanafani, K.; Zimmermann, R.K.; Rasmussen, F.N.; Birgisdóttir, H. Learnings from developing a context-specific LCA tool for buildings—The case of lcabyg 4. *Sustainability* **2021**, *13*, 1508. [CrossRef]
3. Frischknecht, R.; Balouktsi, M.; Lützkendorf, T.; Aumann, A.; Birgisdottir, H.; Ruse, E.G.; Hollberg, A.; Kuittinen, M.; Lavagna, M.; Lupišek, A.; et al. Environmental benchmarks for buildings: Needs, challenges and solutions—71st LCA forum, Swiss Federal Institute of Technology, Zürich, 18 June 2019. *Int. J. Life Cycle Assess.* **2019**, *24*, 2272–2280. [CrossRef]
4. Soust-Verdaguer, B.; Llatas, C.; García-Martínez, A. Critical review of bim-based LCA method to buildings. *Energy Build.* **2017**, *136*, 110–120. [CrossRef]
5. Rasmussen, F.N.; Malmqvist, T.; Birgisdóttir, H. Drivers, barriers and development needs for LCA in the Nordic building sector—A survey among professionals. *IOP Conf. Ser. Earth Environ. Sci.* **2020**, *588*. [CrossRef]
6. Balouktsi, M.; Lützkendorf, T.; Röck, M.; Passer, A.; Reisinger, T.; Frischknecht, R. Survey results on acceptance and use of Life Cycle Assessment among designers in world regions: IEA EBC Annex 72. *IOP Conf. Ser. Earth Environ. Sci.* **2020**, *588*. [CrossRef]
7. Obrecht, T.P.; Röck, M.; Hoxha, E.; Passer, A. BIM and LCA Integration: A Systematic Literature Review. *Sustainability* **2020**, *12*, 5534. [CrossRef]
8. Wastiels, L.; Decuypere, R. Identification and comparison of LCA-BIM integration strategies. *IOP Conf. Ser. Earth Environ. Sci.* **2019**, *323*. [CrossRef]
9. Chastas, P.; Theodosiou, T.; Kontoleon, K.J.; Bikas, D. Normalising and assessing carbon emissions in the building sector: A review on the embodied CO_2 emissions of residential buildings. *Build. Environ.* **2018**, *130*, 212–226. [CrossRef]
10. Russell-Smith, S.V.; Lepech, M.D.; Fruchter, R.; Littman, A. Impact of progressive sustainable target value assessment on building design decisions. *Build. Environ.* **2015**, *85*, 52–60. [CrossRef]
11. Rasmussen, F.N.; Ganassali, S.; Zimmermann, R.K.; Birgisdóttir, H. LCA benchmarks for residential buildings in Northern Italy and Denmark. *Build. Res. Inf.* **2019**, *47*, 833–849. [CrossRef]
12. Blengini, G.A.; Di Carlo, T. The changing role of life cycle phases, subsystems and materials in the LCA of low energy buildings. *Energy Build.* **2010**, *42*, 869–880. [CrossRef]

13. Röck, M.; Ruschi Mendes Saade, M.; Balouktsi, M.; Nygaard, F.; Birgisdottir, H.; Frischknecht, R.; Habert, G.; Lützkendorf, T. Embodied GHG emissions of buildings—The hidden challenge for e ff ective climate change mitigation. *Appl. Energy* **2019**, *258*, 114107. [CrossRef]
14. Moncaster, A.M.; Rasmussen, F.N.; Malmqvist, T.; Houlihan Wiberg, A.; Birgisdottir, H. Widening understanding of low embodied impact buildings: Results and recommendations from 80 multi-national quantitative and qualitative case studies. *J. Clean. Prod.* **2019**, *235*, 378–393. [CrossRef]
15. International Organization for Standardization. *ISO 14040: Environmental Management—Life Cycle Assessment—Principles and Framework*; ISO: Geneva, Switzerland, 2006.
16. International Organization for Standardization. *ISO 14044: Environmental Management—Life Cycle Assessment—Requirements and Guidelines*; ISO: Geneva, Switzerland, 2006.
17. European Committee for Standardization. *EN 15978: Sustainability of Construction Works—Assessment of Environmental Performance of Buildings—Calculation Method*; CEN: Brussels, Belgium, 2011.
18. Lützkendorf, T. Assessing the environmental performance of buildings: Trends, lessons and tensions. *Build. Res. Inf.* **2017**, *3218*, 1–21. [CrossRef]
19. Zimmermann, R.K.; Skjelmose, O.; Jensen, K.G.; Jensen, K.K.; Birgisdottir, H. Categorizing Building Certification Systems According to the Definition of Sustainable Building. *IOP Conf. Ser. Mater. Sci. Eng.* **2019**, *471*. [CrossRef]
20. The government (the Social Democratic Party); Venstre; Dansk Folkeparti; the Socialist People's Party; Radikale Venstre; Enhedslisten; the Conservative People's Party and the Alternative. *National Strategi for Bæredygtigt Byggeri*; The Danish parliament: Copenhagen, Denmark, 2021; pp. 1–8.
21. Meex, E.; Hollberg, A.; Knapen, E.; Hildebrand, L.; Verbeeck, G. Requirements for applying LCA-based environmental impact assessment tools in the early stages of building design. *Build. Environ.* **2018**, *133*, 228–236. [CrossRef]
22. Lamé, G.; Leroy, Y.; Yannou, B. Ecodesign tools in the construction sector: Analyzing usage inadequacies with designers' needs. *J. Clean. Prod.* **2017**, *148*, 60–72. [CrossRef]
23. Saunders, C.L.; Landis, A.E.; Mecca, L.P.; Jones, A.K.; Schaefer, L.A.; Bilec, M.M. Analyzing the practice of life cycle assessment: Focus on the building sector Saunders et al. Analyzing the practice of life cycle assessment. *J. Ind. Ecol.* **2013**, *17*, 777–788. [CrossRef]
24. Marsh, R.; Rasmussen, F.N.; Birgisdottir, H. *Embodied Carbon Tools for Architects and Clients Early in the Design Process*; Pomponi, F.A., Moncaster, C.D.W., Eds.; Springer: Berlin/Heidelberg, Germany, 2018; ISBN 978-3-319-72796-7.
25. Basbagill, J.; Flager, F.; Lepech, M.; Fischer, M. Application of life-cycle assessment to early stage building design for reduced embodied environmental impacts. *Build. Environ.* **2013**, *60*, 81–92. [CrossRef]
26. John, V. Derivation of Reliable Simplification Strategies for the Comparative LCA of Individual and "Typical" Newly Built Swiss Apartment Buildings. Ph.D. Thesis, ETH, Zürich, Switzerland, 2012.
27. The European Parliament and the Council of the European Union. Directive 2014/24/EU of The European Parliament and of The Council of 26 February 2014 on Public Procurement and Repealing Directive 2004/18/EC (Text with EEA Relevance). *Off. J. Eur. Union* **2014**, *94*, 65–242.
28. Erhvervsministeriet. *Udbudsloven*; Erhvervsministeriet: Copenhagen, Denmark, 2015; pp. 1–9.
29. Transport- og Boligministeriet. *Bekendtgørelse om Anvendelse af Informations- og Kommunikationsteknologi (IKT) i Offentligt Byggeri*; Transport- og Boligministeriet: Copenhagen, Denmark, 2013.
30. Laakso, M.; Kiviniemi, A. The IFC standard—A review of history, development, and standardization. *Electron. J. Inf. Technol. Constr.* **2012**, *17*, 134–161.
31. Theißen, S.; Höper, J.; Drzymalla, J.; Wimmer, R.; Markova, S.; Meins-Becker, A.; Lambertz, M. Using Open BIM and IFC to Enable a Comprehensive Consideration of Building Services within a Whole-Building LCA. *Sustainability* **2020**, *12*, 5644. [CrossRef]
32. Horn, R.; Ebertshäuser, S.; Di Bari, R.; Jorgji, O.; Traunspurger, R.; von Both, P. The BIM2LCA approach: An industry foundation classes (IFC)-based interface to integrate life cycle assessment in integral planning. *Sustainability* **2020**, *12*, 6558. [CrossRef]
33. Santos, R.; Costa, A.A.; Silvestre, J.D.; Pyl, L. Integration of LCA and LCC analysis within a BIM-based environment. *Autom. Constr.* **2019**, *103*, 127–149. [CrossRef]
34. Figl, H.; Ilg, M.; Battisti, K. 6D BIM-Terminal: Missing Link for the design of CO-neutral buildings. *IOP Conf. Ser. Earth Environ. Sci.* **2019**, *323*. [CrossRef]
35. Saridaki, M.; Psarra, M.; Haugbølle, K. Implementing life-cycle costing: Data integration between design models and cost calculations. *J. Inf. Technol. Constr.* **2019**, *24*, 14–32.
36. Kiss, B.; Röck, M.; Passer, A.; Szalay, Z. A cross-platform modular framework for building Life Cycle Assessment. *IOP Conf. Ser. Earth Environ. Sci.* **2019**, *323*. [CrossRef]
37. Röck, M.; Hollberg, A.; Habert, G.; Passer, A. LCA and BIM: Visualization of environmental potentials in building construction at early design stages. *Build. Environ.* **2018**, *140*, 153–161. [CrossRef]
38. Tsikos, M.; Negendahl, K. Sustainable Design with Respect to LCA Using Parametric Design and BIM Tools. In Proceedings of the World Sustainable Built Environment Conference 2017, Wan Chai, Hong Kong, 5–7 June 2017.
39. CSTB EveBIM. Available online: https://www.evebim.fr/ (accessed on 23 April 2021).
40. Cavalliere, C.; Dell'Osso, G.R.; Pierucci, A.; Iannone, F. Life cycle assessment data structure for building information modelling. *J. Clean. Prod.* **2018**, *199*, 193–204. [CrossRef]

41. Mora, T.D.; Bolzonello, E.; Cavalliere, C.; Peron, F. Key parameters featuring bim-lca integration in buildings: A practical review of the current trends. *Sustainability* **2020**, *12*, 7182. [CrossRef]
42. Antón, L.Á.; Díaz, J. Integration of life cycle assessment in a BIM environment. *Procedia Eng.* **2014**, *85*, 26–32. [CrossRef]
43. Díaz, J.; Antön, L.Á. Sustainable construction approach through integration of LCA and BIM tools. In Proceedings of the 2014 International Conference on Computing in Civil and Building Engineering, Orlando, FL, USA, 23–25 June 2014; pp. 283–290. [CrossRef]
44. Buildingsmart. Available online: https://www.buildingsmart.org/ (accessed on 11 March 2021).
45. Toth, B.; Janssen, P.; Stouffs, R.; Chaszar, A.; Boeykens, S. Custom digital workflows: A new framework for design analysis integration. *Int. J. Archit. Comput.* **2012**, *10*, 481–500. [CrossRef]
46. Negendahl, K. Building performance simulation in the early design stage: An introduction to integrated dynamic models. *Autom. Constr.* **2015**, *54*, 39–53. [CrossRef]
47. Grasshopper. Available online: https://www.grasshopper3d.com/ (accessed on 15 March 2021).
48. Dynamo. Available online: https://dynamobim.org/ (accessed on 15 March 2021).
49. Cavalliere, C.; Habert, G.; Dell'Osso, G.R.; Hollberg, A. Continuous BIM-based assessment of embodied environmental impacts throughout the design process. *J. Clean. Prod.* **2019**, *211*, 941–952. [CrossRef]
50. Dupuis, M.; April, A.; Lesage, P.; Forgues, D. Method to Enable LCA Analysis through Each Level of Development of a BIM Model. *Procedia Eng.* **2017**, *196*, 857–863. [CrossRef]
51. Durão, V.; Costa, A.A.; Silvestre, J.D.; Mateus, R.; De Brito, J. Integration of environmental life cycle information in BIM objects according with the level of development. *IOP Conf. Ser. Earth Environ. Sci.* **2019**, *225*. [CrossRef]
52. Durão, V.; Costa, A.A.; Silvestre, J.D.; Mateus, R.; Santos, R.; Brito, J. De Current Opportunities and Challenges in the Incorporation of the LCA Method in BIM. *Open Constr. Build. Technol. J.* **2020**, *14*, 336–349. [CrossRef]
53. Frivillig Bæredygtighedsklasse. Available online: https://baeredygtighedsklasse.dk/ (accessed on 16 March 2021).
54. Green Building Council Denmark. Available online: https://www.dk-gbc.dk/english/ (accessed on 16 March 2021).
55. Ökobaudat. Available online: https://www.oekobaudat.de/ (accessed on 16 March 2021).
56. Birgisdottir, H.; Rasmussen, F.N. Development of LCAbyg: A national Life Cycle Assessment tool for buildings in Denmark. *IOP Conf. Ser. Earth Environ. Sci.* **2019**, *290*. [CrossRef]
57. Kanafani, K.; Zimmermann, R.K.; Rasmussen, F.N.; Birgisdóttir, H. Early Design Stage Building LCA Using the LCAbyg Tool: New Strategies for Bridging the Data Gap. *IOP Conf. Ser. Earth Environ. Sci.* **2019**, *323*. [CrossRef]
58. Zimmermann, R.K.; Kanafani, K.; Nygaard Rasmussen, F.; Birgisdóttir, H. Early Design Stage Building LCA using the LCAbyg tool: Comparing Cases for Early Stage and Detailed LCA Approaches. *IOP Conf. Ser. Earth Environ. Sci.* **2019**, *323*. [CrossRef]
59. Theißen, S.; Höper, J.; Wimmer, R.; Zibell, M.; Meins-Becker, A.; Rössig, S.; Goitowski, S.; Lambertz, M. BIM integrated automation of whole building life cycle assessment using German LCA data base ÖKOBAUDAT and Industry Foundation Classes. *IOP Conf. Ser. Earth Environ. Sci.* **2020**, *588*. [CrossRef]
60. Ligurgo, V.; Philippette, T.; Fastrez, P.; Collard, A.S.; Jacques, J. A Method Combining Deductive and Inductive Principles to Define Work-Related Digital Media Literacy Competences. *Commun. Comput. Inf. Sci.* **2018**, *810*, 245–254. [CrossRef]
61. Realdania. *Byggeriets Nøgletal*; Realdania: Copenhagen, Denmark, 2016.
62. Danske Arkitektvirksomheder. Available online: https://www.danskeark.dk/content/tal-analyser (accessed on 16 March 2021).
63. Röck, M.; Passer, A.; Ramon, D.; Allacker, K. The coupling of BIM and LCA—Challenges identified through case study implementation. In *Life Cycle Analysis and Assessment in Civil Engineering: Towards an Integrated Vision*; Taylor & Francis Group: London, UK, 2019; pp. 841–846.
64. Petrova, E.A.; Romanska, I.; Stamenlov, M.; Svidt, K.; Jensen, R.L. Development of an Information Delivery Manual for Early Stage BIM-based Energy Performance Assessment and Code Compliance as a Part of DGNB Pre-certification. In Proceedings of the 15th IBPSA Conference: Building Simulation 2017, San Francisco, CA, USA, 7–9 August 2017. [CrossRef]

Review

The Role of the Interface and Interface Management in the Optimization of BIM Multi-Model Applications: A Review

Nawal Abdunasseer Hmidah [1,*], Nuzul Azam Haron [1], Aidi Hizami Alias [1], Teik Hua Law [1], Abubaker Basheer Abdalwhab Altohami [1] and Raja Ahmad Azmeer Raja Ahmad Effendi [2]

1 Department of Civil Engineering, Faculty of Engineering, University Putra Malaysia (UPM), Serdang 43400, Malaysia; nuzul@upm.edu.my (N.A.H.); aidihizami@upm.edu.my (A.H.A.); lawteik@upm.edu.my (T.H.L.); gs50517@student.upm.edu.my (A.B.A.A.)

2 Department of Industrial Design, Faculty of Design & Architecture, University Putra Malaysia (UPM), Serdang 43400, Malaysia; azmeer@upm.edu.my

* Correspondence: gs57074@student.upm.edu.my or nawal.aljahmi@yahoo.com

Abstract: This review targets the BIM interface, the BIM multi-model approach, and the role of employing algorithms in BIM optimization to introduce the need for automation in the BIM technique, instead of complicating manual procedures in order to reduce possible errors. The challenge with adopting BIM lies in the limiting ability of computer-aided design (CAD) to generate a read-able and straightforward Revit by BIM, requiring the homogeneous data format to be generalized better and maintain a super data mod. Furthermore, the communication and management inter-face (CMI) faces some shortcomings due to limitations in its ability to recognize the role of the interface during the project construction phase. This review demonstrates several proposals to simplify the interface, in order to facilitate better communication amongst participants. The industry foundation class (IFC) model requires a new technique to unlock the potential future of intelligent buildings using the BIM multi-model approach integrated with the Internet of Things (IoT). Trials conducted to enhance the BIM model lack advanced methods for optimizing cost, energy consumption, labor, material movement, and the size of layout of the project, by utilizing heuristic, metaheuristic, and k-mean algorithms. The enhancement of BIM could involve algorithms to achieve better productivity, safety, cost, time, and construction frameworks. The review shows that some gaps and limitations still exist, especially considering the potential link between BIM and building management system (BMS) and the level of influence of the BIM-IoT prototype. Future work should find the best approach to solve facility management within the dynamic model, which is still under investigation.

Keywords: BIM; management interface; BIM multi-model; BIM-BMS system; optimization

Citation: Hmidah, N.A.; Haron, N.A.; Alias, A.H.; Law, T.H.; Altohami, A.B.A.; Effendi, R.A.A.R.A. The Role of the Interface and Interface Management in the Optimization of BIM Multi-Model Applications: A Review. *Sustainability* **2022**, *14*, 1869. https://doi.org/10.3390/su14031869

Academic Editor: Antonio Garcia-Martinez

Received: 9 September 2021
Accepted: 21 November 2021
Published: 7 February 2022

1. Introduction

BIM is a project-improving tool that globally provides a revolutionary platform for design, construction, maintenance, operation, and improvement in various fields for the rehabilitation, retrofit, and redevelopment of existing assets in the built environment. Another helpful definition considers BIM as a methodology that combines several processes and tools to improve projects and overall construction outcomes [1].

BIM is a paradigm that shifts the inefficient 2D drawing and processes and practices of documentation towards much more precise model-centric processes and practices. Researchers consider BIM as an integrated information system that effectively assimilates the organizational functions and processes of project delivery. BIM is an inclusive term that can be defined in diverse ways; however, the most typical definition states that BIM is software used to create value and promote collaboration in the entire lifecycle of an asset, using underpinning theories by collating and exchanging 3D models [2].

BIM has reached an exciting stage, as many built environment stakeholders are currently using or considering using it. As reported in 2017, 86% of UK respondents expect to adopt BIM for their projects [3]. BIM adoption level varies from one country to another, depending on the size and complexity of the projects.

At its inception, building information modelling (BIM) was associated with using 3D modeling with the availability of various software tools and techniques. Traditionally, although 3D construction models had been integrated for additional measurements of time and cost, they were found inadequate in terms of including all the project-specific details necessary for a building project. On the other hand, BIM is equipped with enough technicalities to create virtual 3D models by integrating relevant information, and simultaneously granting project participants a better understanding of the project phases [4]. Concerning expanding models, Ivson et al. [5] developed several models that were utilized to represent models and several corresponding sub-models to serve different operations simultaneously.

Accordingly, the philosophy of the multi-model approach has emerged to collate data from various sources with different formats into a single exchangeable resource [6], which can be characterized as object-oriented [7].

The objective of BIM is to create accurate, reliable, complementary, and replaceable information for the construction of buildings [8]. These objectives can only be attained by implementing interoperability and parametric (adjusting variables) behavior. Eastman et al. [9] defined BIM as a technology equipped with a set of processes that aim at producing, communicating, and analyzing building models. BIM is widely considered a significant factor in the construction industry. BIM describes an integrated model-based view of a facility's lifecycle, including design, planning, and construction, as well as operation and maintenance (O&M) [10].

Recently, BIM has been adopted in various types of projects, and in projects that require dynamic data exchange amongst multiple actors with information aggregation, such as designing a project, running software, handling data, revising all or parts of the project [11], and improving the efficiency of construction [12].

Briefly, the BIM model is described as a mixture of graphical and non-graphic data that can communicate throughout specific data-exchange formats.

Recently, it has been observed that BIM applications are expanding to many fields, owing to the introduction of 3D geometric models and 3D coordination [13]. These applications go beyond architecture and engineering, to cover and initiate a strong motive for homeowners, facility managers, contractors, and fabricators [14]. The project focuses on BIM adoption, provided by utilizing automation in the modeling process. This modeling improves communication and accuracy among various parties throughout, exchanging views and reducing the errors in the coordination of building activities [15]. BIM applications plan, design, build, construct, operate, and reduce energy consumption [16]. These developments were not applicable to certain countries; they are, instead, applied to all countries, sharing the same principles of integration of BIM and building energy management (BEM) in a single tool [17].

The fundamental contributions of BIM are in energy-related matters, simulations, and information, which can be described as involving the automation of energy to better present output in order to enhance storage and organizational capabilities concerning new-building data. The other contribution of BIM concerns the facilitation of output presentations in energy management systems [18]. Similarly, [19] studied a conceptual framework for a BIM-based energy management support system (BIM-EMSS) by developing a real-time energy simulation using eQuest. Based on the determination of [18], visualizing the geometric data in BIM could allow the user to monitor the real-time energy performance of different zones in a building.

Other benefits of BIM applications include storing, monitoring, and organizing energy-related information in real-time energy systems. The system can generate information related to home energy consumption and how to relate activities to environmental temperature and the degree of occupancy. The adoption of BIM models in real-time energy monitoring systems was explained by Alahmad et al. [20], who proposed a combined system that uses a hardware component system and a software system. Woo et al. [21] reported other BIM applications in a building equipped with sensors that provide real-time data to BIM models using a standard schema to facilitate processing the data related to sensors and actuators. The other important application of BIM is performed by linking existing libraries, where a great amount of information about the thermal conductivity properties is available. The life cycle assessment of a building can be estimated better by integrating CAD and BIM. This link provides information about optimizing the building envelope or sizing the HVAC system [22].

2. BIM and Interface Management Flowchart

The flow of topics in this review is outlined in Figure 1. The purpose of this flowchart is to guide readers for easy access to the topics that are included in this review, and to present the contents in a structured fashion.

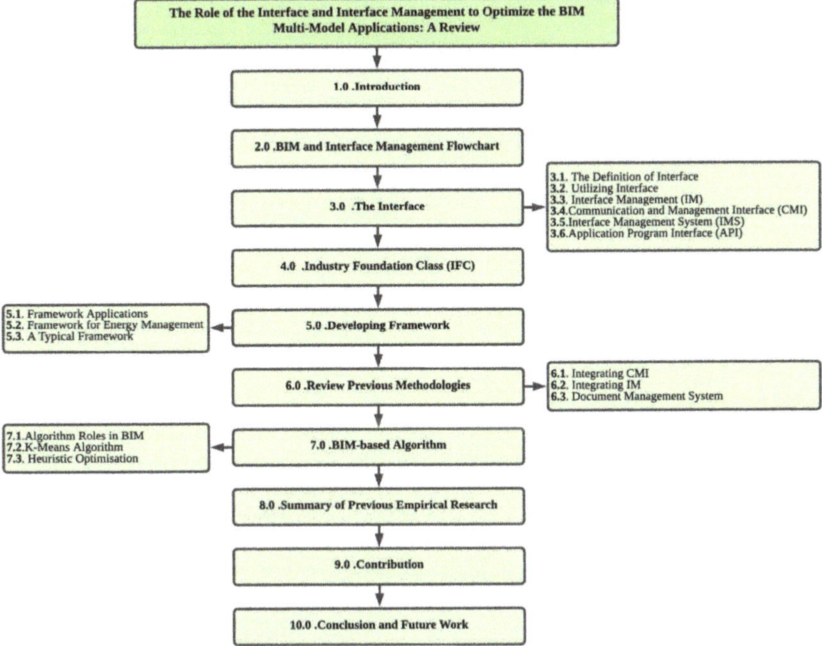

Figure 1. Contents flowchart.

3. The Interface

3.1. The Definition of Interface

The definition of the interface has been developing since 1967, when Wren [23] denoted that the interface is the contact point or set of points (surface) between two independent systems to achieve a better, more extensive and unified system. However, as time progresses, other researchers have been proposing numerous definitions of the interface.

Lin [24] expanded the role of the interface to include cases related to different opinions, such as schedule, cost, technical areas, and the space between systems. Interface management (IM) is another element of interface to address the challenges of managing complex capital projects to face the rising complexity due to globalization and the geographical distribution of various cultures [25]. Shen et al. [26] have provided another depth of the interface by correlating the interrelation and interaction among different organizations and stakeholders. Profoundly, the interface helps organizations to eliminate the loss of information and leverage the data in BIM models to improve communication and collaboration between architects, engineers, contractors, and facility managers. Recently, a definition of the interface has been adopted in a broader scope to include all common boundaries and non-physical interaction between systems, organizations, stakeholders, project phases and scopes, and construction elements [27]. The interface is a virtual entity whose aim is to help the organization by eliminating information loss. In addition, it improves communication and collaboration between various stakeholders, such as architects, engineers, contractors, and facility managers. Hence, data are supposed to seamlessly transfer through the interface domain between designs, construction, operations, and maintenance [28]. Table 1 details the features of the various interface types.

Table 1. Characteristics of the interface types.

Reference	Interface Type	Details
[29]	Physical	Physical connections between two or more elements of the building or components.
	Contractual	Work packages associated with specialist contractors.
	Organizational	Lifecycle relationship between parties involved in the project.
[30]	Intrinsic	Physical links among the various components.
	Discipline	Theknowledge areas that are necessary to engineer develop studies, analyses, designs, sufficient to utilize the concept.
	Project	Strategies among contractors, subcontractors, vendors and any external provider.
[31]	Functional	All sub-functions activities and components.
	Physical	Interfaces between physical sub-systems.
	Hieratical	Between top and low organizational segments regarding project objectives.

3.2. Utilizing Interface

Recording information belonging to managing complaints and responses using emails is not good for solving interface problems. Recently, researchers have developed two critical BIM and interface management (MI) approaches for managing more complex projects [32]. Originally, IM was used as an information-intensive task to provide helpful information to participants [33]. Meanwhile, IM is currently recognized as the most critical organizational strategy in construction management [34]. One of the reasons that made IM an emerging construction strategy was the ability to resolve and enhance construction management by tracking, managing, and eliminating unnecessary mistakes [28]. Hence, project members can locate current interfaces to work out any existing interface issues. It is noted in current construction that without IM implementation, the project could experience design errors, a component malfunction, device performance failures, organized difficulties, and construction disputes [35]. Sacks et al. [14] have emphasized that the BIM models provide a natural interface equipped with sensors and remote FM operations to support monitoring and control practices. Applying BIM in construction management helps project stakeholders monitoring, handling, and tracking all issues relevant to 3D modeling. However, the present BIM-based information systems are still using IM because of the lack of reliable IM communication and management of interfaces (CMI) in the BIM environment [36]. Regarding the history of the construction, the maps of the 3D interface provide information that belongs to the past, present, and future interactions, highlighting an overview of real-time project history.

In construction, interfaces, without distinctive categorization, are either internal (emphasis on contractual relationships) or external (contracts or scopes of work). Based on this characterization, internal interfaces are easier to handle, because they deal with only one team rather than two or more teams, as in the external interfaces. However, when the number of contractors is large, managing interfaces becomes very difficult, and, as such, it is important to conduct certain classifications between contractors and subcontractors based on the given responsibility [37].

3.3. Interface Management (IM)

In the late 1960s and early 1970s, interface management (IM) was introduced to ensure matching the specification of the interface system, data, and missing equipment [38]. Later, IM was used to identify organizational, managerial, and technological interfaces throughout establishing interrelationships [38]. To the best of the authors' knowledge, IM was not wholly integrated into engineering and construction procedures, because of lacking technical infrastructure. In achieving such a goal, several organizations have established IM groups inside their management practices, playing an essential role in training employees to better understand the role of interface manager and interface coordinator.

Interfaces emerged in splitting a project into many sub-projects carried out by several soft or hard, external or internal entities [39]. The soft interface can exchange design criteria, clearance requirements, or utility requirements between the engineering and delivery team and an external party. On the other side, complex interfaces that deal with physical connections between two or more components or systems are examples of hard interfaces. These interfaces include structural steel connectors, pipe terminations, and cable connectors. The interface management process aims to enter into agreements with other stakeholders about roles and duties, time to provide interface information, and early identification of primary interfaces [40]. Having a defined method for exchanging information enables detailed monitoring of performance in meeting requirements, with any inadequacies identified and remedied quickly. Furthermore, interfaces are classified into several groups to fulfill specific goals, such as organizational split interface [29] or resource interface [41].

Figure 2 contains the four main components of the IM system, including the interface of stakeholders, interface points (IPs), interface agreements (IAs), and interface agreement deliverables (IADs). Meanwhile, the interface action item (IAI) consists of tasks and activities aiming to facilitate the agreement of the four IA components between stakeholders. The task and activities are deliverable by IAI to perform an interface agreement (IA), which combines all activities, such as scheduling, drawings, quotations, and evaluation.

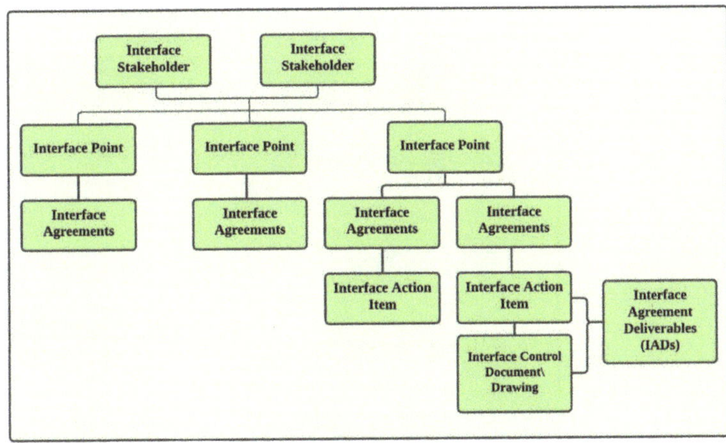

Figure 2. Components interface management [27].

3.4. Communication and Management Interface (CMI)

The format limitation of the standard BIM file-based model could be used to share the most recent building progress [41]. CMI was integrated into BIM to facilitate discussing, sharing, and responding to issues related to the BIM elemental interface during the construction phase [28]. CMI enables project engineers and managers to access previous records regarding BIM models for a given project. In the future, it will manage the response to interface problems, as illustrated in Figure 3. The literature focuses on CMI integration; however, it lacks a suitable platform for BIM-based CMI [42].

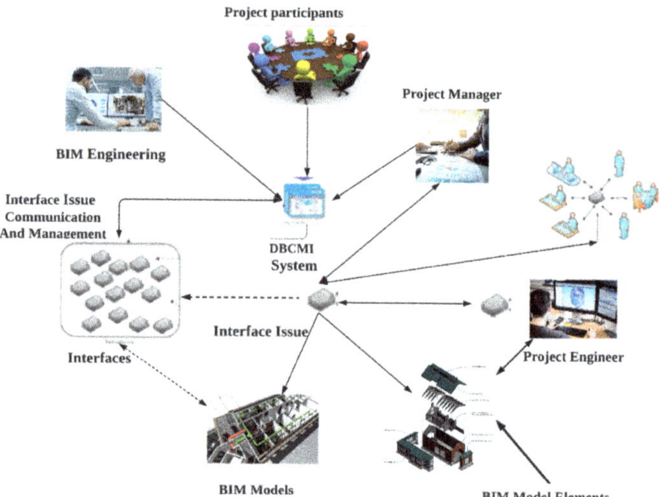

Figure 3. Application of integrated CMI in BIM for construction interface management [28].

3.5. *Interface Management System (IMS)*

In 2014, IMS was defined, within the guidelines of interface and IM, as a combination of managerial and relational communication that can be delivered among two or more interface stakeholders. [43]. It was mentioned earlier that IPs, IAs, and IADs are the elements of IMS. IMS was studied in terms of a six-step framework execution, as shown in the self-explanatory Figure 4.

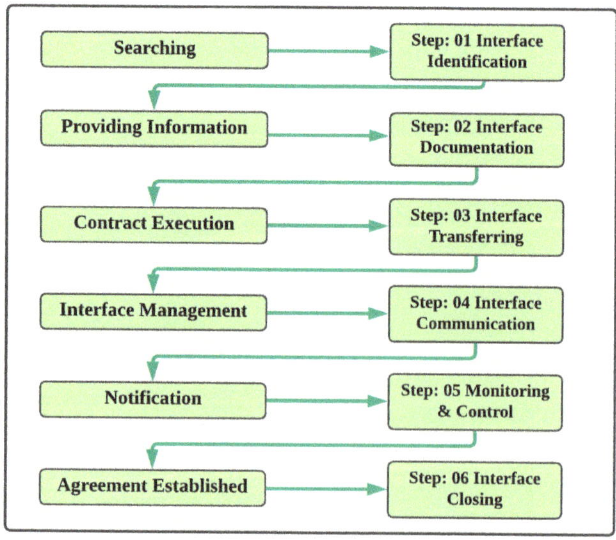

Figure 4. Mechanism of IMS framework from searching to the end of the contract.

Recently, there has been increasing concern about IM practices among contractors. Based on the IM definition, a new framework could be used to define the interface management system (IMS). IMS includes many IPs, with each IP including multiple IAs, and each IA may include various IADs [27]. There are several types of interface management, as explained in Table 2.

Table 2. Characteristics of the IM interfaces [25].

Category	Definition/Purpose
Interface Management (IM)	Managing relational communications between more than one stakeholders.
Interface Stakeholder	It is a part of formal interface management agreement of the project.
Interface/Interface Point (IP)	It is the soft (hard) contact point between two interdependent interface stakeholders.
Interface Agreement (IA)	The formal communication documents between two interface stakeholders concerning desription, actions involved, and dates.
Interface Action Items (IAI)	IAI regulates tasks and activities to perform the defined agreement in each interface agreement.
Interface Control Document Drawing (ICD)	To identify and capture interface information prior to approvement. ICDs are useful for separate organizations with a common particular interface.

Based on the above discussion, interface stakeholders are involved in many deliverable information or tasks to handle the interface efficiently. In each interface point, there are numerous interface agreements. These agreements can be delivered to other parties. Each interface stakeholder can deal with several interface points and agreements, as illustrated in Figure 5.

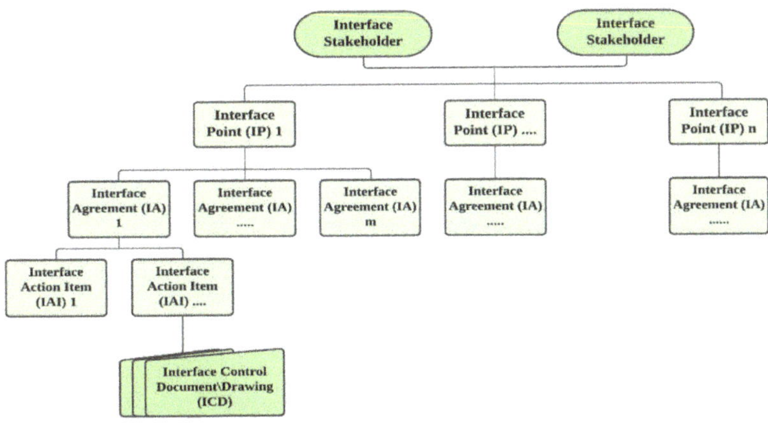

Figure 5. Hierarchy of Interface Management Elements [25].

3.6. *Application Program Interface (API)*

API was defined in 1968 as a collection of code routines to provide external users with data and data functionality that was used in programming libraries [44]. API was also used interchangeably with frameworks, libraries, and operating systems. Nevertheless, the current colloquial definition of API refers most typically to a synonym for web API [45]. According to Programmable Web-based API directory, the number of available APIs continues to expand, particularly those classified as data, financial, or analytics [5]. Numerous accessible APIs enable access to massive data volumes. APIs are versatile technical solutions that may be utilized in various applications. The first application is the Google Maps Platform, which has a Places API that provides access to over 150 million locations worldwide. Firms use APIs to refer to products, add more data to databases, or create specific APIs [46]. The second application is to perform functions related to procedural languages, such as C, to act as a function call, by involving information about all the functions and routines that it provides.

APIs are a collection of methods that enable programs to access data and communicate with external software components, operating systems, or microservices. APIs are a critical component of many modern software architectures, because they provide high-level abstractions that simplify programming processes, create distributed and modular software systems, and allow code reuse [47]. Hence, APIs make necessary accessible functionalities for developers to enable IoT cloud infrastructures [48]. APIs are digital apps that can help in communicating with back-end services [45].

AP can be expanded to describe all calls, subroutines, or software, to enable application programs in services such as application, operating system, network, or another lower-level software program [49]. APIs facilitate information for developers to work with essential capabilities or data to leverage and govern IoT cloud infrastructures. APIs allow partners or the public to activate participants and generate new revenue streams [50].

Additionally, APIs help establish an interface that connects functions in one unique system, cutting down transaction costs, and improving efficiency [51]. Another feature of API is providing third-party developers with access to private data owned by Google, Facebook, Twitter, and many other large firms [52].

APIs, then, constitute the interfaces of the various building blocks that a developer needs to create an application [53]. An API currently summarizes a set of programming codes to transfer data between one software product and another. APIs are composed of two fundamental components: technical specifications describing data exchange choices between solutions in the form of a request for processing, and data delivery protocols and a program interface based on the specifications they represent. Today, APIs come in three types—standard, widespread, and versatile. Figure 6 explains the main types of API and the corresponding policies.

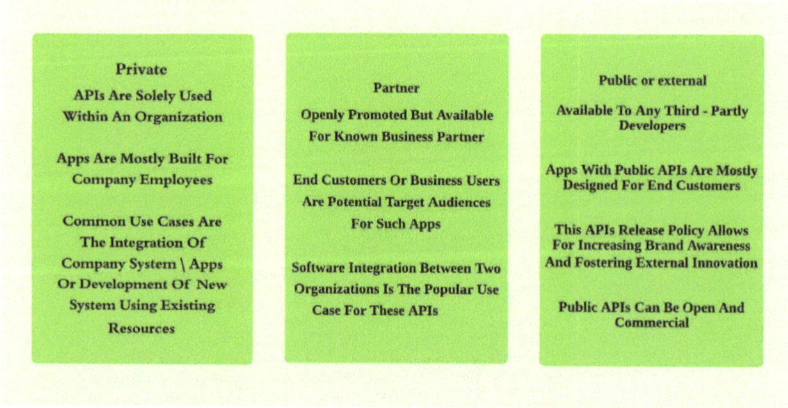

Figure 6. API types and functions [51].

4. Industry Foundation Class (IFC)

Sheng Jun et al. [54] proposed several methods to transfer various software programs and data formats using IFC or DWG. There are many approaches to co-ordinate IFC, IDM, Open BIM Collaboration Format (BCF), Open BIM Collaboration Format (BCF), and Model View Definition (MVD). For instance, the formats of BIM can open communication between two users based on the IFC standard data model. IFC is equipped with a high degree of interoperability that can facilitate the opening data standard known as buildingSMART (ISO 16739), depicting the whole building geometry. In addition, IFC provides digital building models to help architects sharing the BIM environment. The Information Delivery Manual (IDM) is a systematic tool for identifying and specifying information flow during a facility's lifecycle. Furthermore, IFC has been developed by buildingSMART and registered under ISO 29481-1:2010 and ISO 29481-2:2010. Concerning IFC schema needed to satisfy one or many ERs, MVD defines a subset published by the software tool ifcDoc developed by BuildingSMART [55]. Furthermore, Zhang [55] considers BCF is an important tool to exchange information in terms of queries, ideas, or demands between different software products, resulting in a technological solution for communication among stakeholders.

In addition to addressing the IFC data model, BCF addresses the position as defined in snapshots or camera perspectives. It is well-known that transforming data into another using software applications with heterogeneous models can be conducted with a single multimodal [7]. Hence, the common practice of BIM models is to exchange information about building structures throughout their life cycle, which is a standard industry practice. The IFC standard in BIM applications acts as a medium for data exchange across domains and stages [56]. The domain of exchanging data for BIM modeling resembles the IFC scheme subset [57]. The benefit of IFC-mapped data exchange is to help the software vendors developing practical import/export features to allow project participants to share and exchange BIM model information. During the different stages of projects, BIM plays an important role in exchanging data and information with specific formats amongst architects, engineers, clients, and contractors to serve throughout the project lifecycle [58].

In construction, stakeholders rely on each other to acquire details. The most critical issue here is to automatically interpret and process the information mapping into data for BIM applications with cross-domain and inter-stage coordination [56]. This process leads to an automated system that needs reliable interoperability for marketing and technological levels [59]. The interoperability process requires exchange information for all contributors to understand the need and provide this information for usage. The goal of interoperability is to provide a better communication system that can be placed at various levels and contribute to achieving the result. Interoperability creates the significant digitalization of the whole process towards full automation and efficient management of these processes [8]. The preparation to establish a construction is a multi-facet issue that started before the initiation of the construction and continued for the whole lifecycle.

One of the most important matters is the scope of each construction, which consumes time with the collaborators throughout the project phases. In this case, there is software to be developed by architects, structural engineers, and designers to store and analyze data. This is called heterogeneous information, since the data are stored, shared, and preserved in different realms to ensure consistency [60].

The construction industry domains involve distinctive advanced data exchange for BIM models, using specialized fields of architecture and construction such as neutral formats found in the industry corporation categories. The IFC schema for diverse disciplines should define the type of BIM standards [61]. IFC, an exchangeable neutral format, is often used in design, engineering, manufacturing, and facility management [62]. The data exchange amongst various software across BIM models and the relevant incremental or "as-built" collection archiving is the main scope of the IFC applications [61].

5. Developing Framework

The framework is defined in different ways, and for various purposes. In the early 1980s, Model-View-Controller (MVC) was the first object-oriented framework [63]. Since then, several papers have been published to show the broad and spread nature of using the term 'framework'. The most reliable definition for the framework is that the one connected to software engineering, which refers to designing and implementing large object-oriented software [4].

However, in a different approach, the framework known as conceptual structure is heavily used to solve or address complex issues containing tools, materials, or components [64]. The framework could be defined as an object-oriented form that can be embodied classes and yielding a solution to a family of problems [65].

Technically, the framework represents a reusable design for a system wholly or partially by setting the abstract classes and their interaction [66]. Another standard definition for the framework depends on formulating a skeleton of a customized application to be suitable to an application developer [67]. Despite the several definitions due to the framework's purpose and nature, the general definition of the framework could include structure, aims, and interrelated parameters within a particular phenomenon. Hence, the framework is a comprehensive architecture that outlines the decomposition of a program into a collection of interacting elements [68].

5.1. Framework Applications

Based on multiple definitions of the framework, it can be helpful for various purposes. Framework, as an application, has become a set of elements for designing or developing a reusable code. The technical difference between a framework and an API is that an API is only a part of a framework. The development framework is being employed to present data integration climate, as corporations seek to decrease cost by outsourcing project-based solutions to temporary staff and third-party firms. Wu and Simmons [69] have confirmed that project planning is vital in the current software development process. Hence, it is necessary to aid in the comprehension and application of the answer [70].

There are two types of frameworks: theoretical and conceptual. This study distinguishes between these two types of the framework, since those two types provide direction and stimulus to study and extend knowledge. According to Grant and Osanloo [71], a theoretical framework is the 'blueprint' or guide for research or providing a specific theory or set of theories concerning a particular area of human effort that may be used to analyze occurrences. By using the theoretical framework, research endeavors will get several benefits. Researchers show how they define the study philosophically, epistemologically, methodologically, and analytically with the addition of the structure [71]. The theoretical framework serves as a guide, and should be consistent with all aspects of the research process, including the formulation of the problem, the review of the literature, the methodology, the presentation, and discussion of the findings, as well as the conclusions derived from them [72].

5.2. Framework for Energy Management

The energy sectors in the construction are a comprehensive and essential part of the construction, which is denoted by the Building Energy Management System (BEMS). Hence, it is expected that integrating BEMS in BIM creates an effective energy data monitoring framework using the human–machine interface (HMI) [73]. Researchers investigated the related technology trends and derived BIM-based HMI framework requirements by identifying the role of each component of the framework. Furthermore, an interface is designed between BIM and BEMS with consideration of HMI, and a well-prepared questionnaire.

According to Siao et al. [74], IM has been recognized as the most critical organizational strategy in construction management, because IM is fundamentally reported as a routine process operation guided by specific control of communications [75]. The need for IM and IMS has become more apparent as the construction becomes more complex, with a considerable increase of participants [41]. The gap inherited from several academic studies was the lack of systematic approaches for managing interfaces during construction and assembly phases [76]. In 2009, Lin [24] had identified four primary interface problems, including insufficient platforms for construction project management, improper managing interface conflicts, problems of managing time, space, and efficiency during the construction phase, lack of an effective mechanism for tracking and managing interfaces, absence of complete official record amongst participants, and difficulty of tracking interface events and obtaining interface information from other participants [33].

The role of IM in tracking all participants' involvement could lead to improving operational management, minimizing detrimental change, and enhancing beneficial change. Morris [38] identified two interfaces—static and dynamic. Furthermore, other interfaces, such as personal, organizational, and system interfaces were identified by [39]. More closely, Pavitt and Gibb [29] proposed three main interface types: physical, contractual, and organizational. The significant number of interfaces could create difficulties in applications. Hence, to simplify this issue, the search for an interface should be aligned with the construction phase. In this phase, interface problems can be categorized as construction, processing, space-related, communication, and variability problems interfaces [33].

The development of 2D to 3D patterns improves the shape, size of a component, and spatial relationships between the components. BIM, as a digital tool, can continue updating and sharing project design information [77]. However, the 3D pattern requires precise geometry to support the design, procurement, fabrication, and construction activities [9]. Accordingly, BIM-based visualization could express information more intuitively by realizing real-time construction [78]. In addition, 3D also provides participants mindful of accuracy and adequacy [79]. BIM and CAD share similar views concerning the construction interface management and develop ConBIMIM system, a mixture of construction, BIM, and IM [33].

BIM is a comprehensive system which enables participants to track project updates and to proide data and information about models whose aim is to manage the effects of the databases on a specific model, capturing information from a particular model, and preserving adding industry-specific applications [80]. IMS, on the other hand, is the source for providing a simple and straightforward representation of various interfaces; clarifying the events of the current interfaces; extending the relationships among interface events, and helpings BIM users to track and identify interface events using different colors [33], as shown in Figure 7.

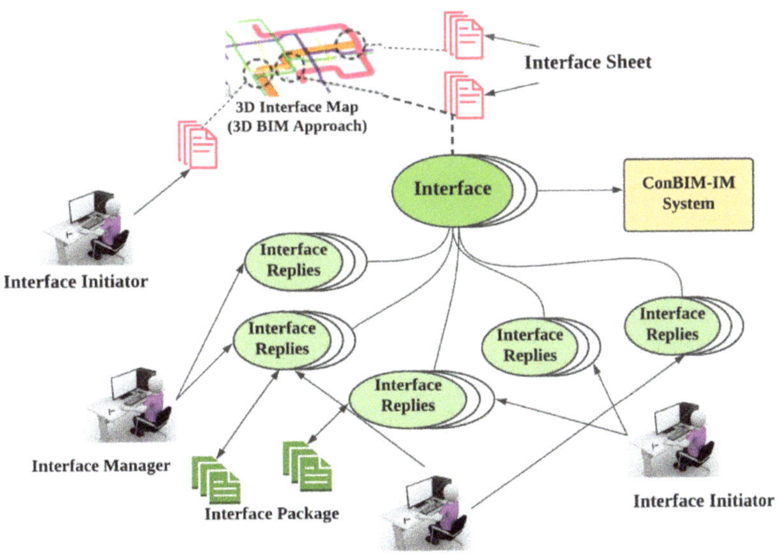

Figure 7. The mechanism of interface management system [33].

The ConBIM-IM system was proposed to design by constructing IM and IMS. Meanwhile, the 3D-CAD interface represents objects and attributes of interface events—the BIM stores digital interface information to facilitate easy updates and interface transferring. As a result, the 3D interface information can be identified, tracked, managed, and further solved problems. The ConBIM-IM enables participating engineers to share and save all documents in 3D formats, and be available upon future request. Figure 8 details the 3D interface maps framework equipped with the eight components of ID, topic, date, description, owner, people, attachments, and history.

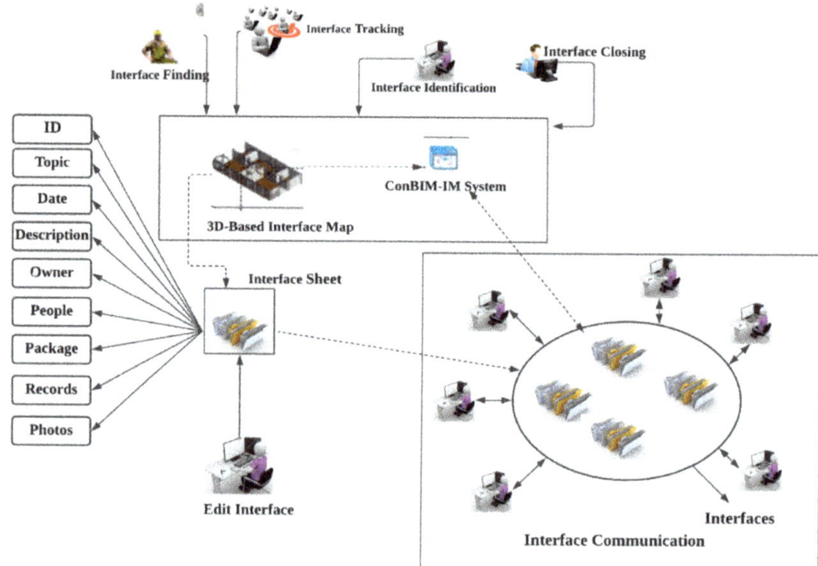

Figure 8. The concept and framework of the 3D-based interface maps approach [33].

5.3. A Typical Framework

Figure 9 shows the BIM-based interfaces framework communication and management integrated with the Interface Breakdown Structure (IBS) and MBS. The process then creates an IBS and a Model Breakdown Structure (MBS) before integrating them in BIM. IBS can break IM into elements of related interfaces. Meanwhile, IBS is a hierarchical representation of interfaces, starting at higher levels, and increasing to more acceptable level interfaces. Furthermore, MBS in interface management is a deliverable-oriented breakdown of a BIM model into more minor elements for interface management. MBS is a crucial interface integrated with elements of BIM models. The CMI-related information stored in elements of the BIM model includes both CMI-related problems and solutions [81]. The CMI essential information should include the interface description, responding, or related attachments such as documents, reports, drawings, and photographs. CMI then enables communication and activates responses associated with projects, activities, people, and organizations. Identifying the connection between the information of CMI and the corresponding interfaces is crucial to the project's management.

In addition to these developments, project engineers can acquire CMI-related issues before sharing them with corresponding BIM model elements. The 3D BIM model known as the DBCMI system can be illustrated at different CMI access levels depending on user roles. As the information is updated in the DBCMI system, the server automatically informs corresponding participants by sending e-mails to the project participants. CMI is equipped with an initial stage through which all responsible participants or project managers are identified. The second stage allows the project participants to edit the information sent and select the appropriate BIM model. In the final stage, the engineers can submit the interface issues associated with the BIM model elements to the DBCMI system for approval. After the approval stage, the corresponding participants respond to problems via the selected interface in the DBCMI system. The system can track all these activities to show the status and the results for each interface problem [81].

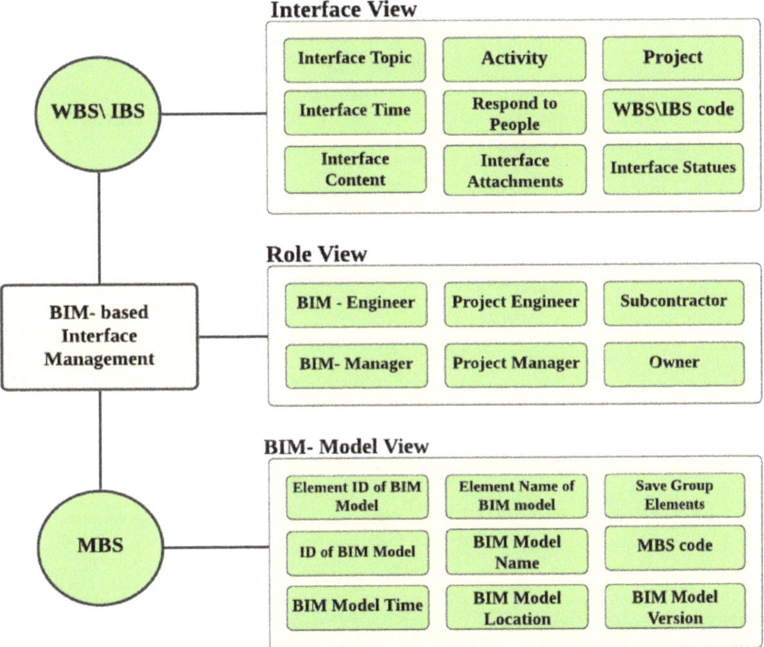

Figure 9. The framework of database- and BIM-based interfaces communication and management [28].

There is another approach to generate a dynamic energy simulation model for a single existing building by collecting existing data to prepare energy retrofits at the lowest cost possible. The proposal includes establishing a polygon model by employing photogrammetrically generated point clouds with the Tool for Energy Analysis and Simulation for Efficient Retrofit (TEASER) and AixLib. A single-family house was taken as a case study to achieve the purpose. The model reproduces the internal air temperatures during synthetical heating up and cooling down, with building heat transfer coefficients (HTC) agreeing within a 12% range. The model requires accurate window characterizations and justifies the use of a very simplified interior geometry. However, uncertainties arose regarding comparing different typologies showing differences in pre-retrofit heat demand of about ±20% to the average [82].

In modern environments, high-rise buildings have become indicative of a diverse building environment that requires special treatment by monitoring activities such as fire hazards. BIM limits fire accidents by creating, developing, and implementing an integrated fire disaster prevention system. The disaster response system is composed of a complete plan, including prevention and evacuation. However, this alarming disaster system is prone to human errors, wrong location, poor communication, and incompleteness. However, the role of BIM is to minimize possible human errors. The system could be better performed in case of providing 3D visualization to support the assessment, planning, and detection of fire safety [83].

6. Review Previous Methodologies

6.1. Integrating CMI

Lin et al. [28] conducted a pilot study to investigate the connection between CMI and MI by interviewing project managers and expert engineers. The study concluded that there was no suitable platform supporting IM, incomplete records for communication and supporting documents, and no clear distinction for how these problems were developed. Hence, CMI must be fully incorporated in construction processes and activities to satisfy these problems, recording all communications at each interface, and linking CMI to BIM models. Lin et al. [28] proposed a concept in which IM was prepared for a general contractor (GC), as shown in Figure 10.

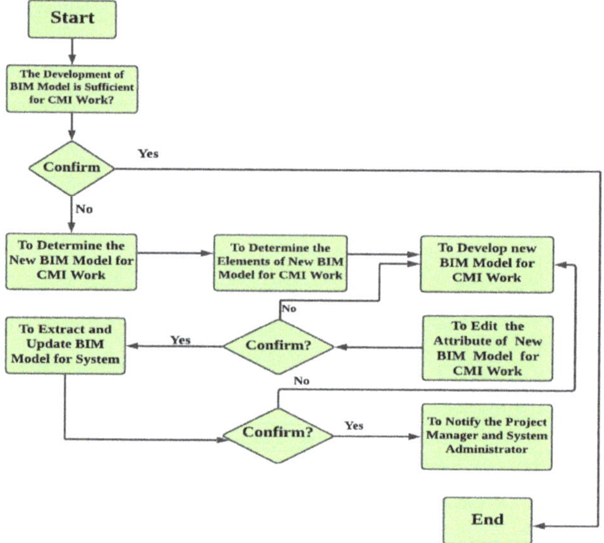

Figure 10. Investigation of CMI and BIM model [28].

6.2. Integrating IM

IM integrating into BIM can be considered a robust approach to improving project monitoring and control to enable real-time decision-making. Most constructions are currently using IM and BIM separately, and, hence, combing IM and BIM is very useful for management, especially for deterministic product management perspective and to a better understanding of managing the complexities, uncertainties, and risk in organizational structure, coordination, collaboration, and communication [32].

The complex construction projects require creating a BIM model before establishing its IM system using a conceptual BIM model that can be generated and detailed during the project lifecycle. On the contrary, an IM system starts in the design phase as part of the project's dynamic systems packages that include changing or evolving elements or removing items from the system. In the construction design phases, many new elements may suffer from repeating, cancellation, or modifying. In all such cases, editing the project essentials has become a necessary step to follow-up the execution of the project. These changes may require updating the project participants as IPs change on the IM system during the project lifecycle. As the project moves from a particular phase to another, the number of participants increases. Then, the number decreases towards the end of the project. This behavior means that IM expands and shrinks, with the change in the number of project participants affecting the number of interface points during the project lifecycle [32].

The IM system consists of interface points (IPs), interface agreements (IAs), and interface agreement deliverables (IADs), as shown in Figure 11. In addition, IPs may contain many IAs, with each IA possibly including many IADs, which means that a typical project may include tens of thousands of IADs which need accurate management to design phases of complex projects that work towards reducing cost and improving [25].

Since 2013, researchers have been steadily developing an effective web-based IM system platform, such as Lin [84], who instrumented a connection amongst project participants for managing interface problems during the construction phase. Within four years, Ju et al. [85] had successfully developed an integrated interface model through which the traditional methods have been changed to a more standardized and structured aiming at improving an IM system. During the period from 2013 until very recently, the heavy research on IM was not successful in eliminating the gaps about visualizing the IM system in the design phase [86].

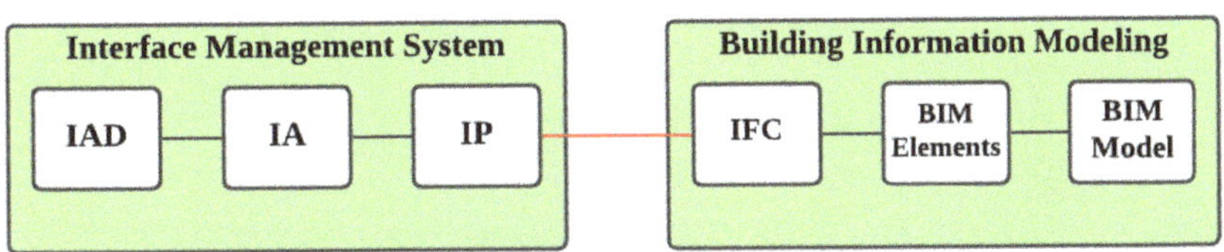

Figure 11. The mechanism of connection IM and BIM [27].

6.3. Document Management System

Complete knowledge about the building material and corresponding repair and condition information is provided in the 'SharePoint' electronic document management system (EDMS). The most crucial benefit of EDMS is that storing information can be done without a formal structure. The information becomes available and can be obtained at any time to be used as needed. Recently, the EDMS information has been stored with some structure to the documents to transfer information quickly. The requirement of the information could become a computable BIM data parameter and to enable reviewing the existing template documents [87].

Regarding the Asset Lifecycle Information Management (ALIM), BIM could benefit from the open standards that enable future builders and developers to record information on the entire lifecycle. ALIM arises because of the availability of information as an essential part of structural digital twins. This type of system provides detailed insights into performance and processes through simulation. This information can be combined and reused quickly if specified, designed, realized, managed, maintained, or dismantled according to open standards. Linking and connecting data are innovative concepts, since ALIM requires bringing together diverse sources from different domains with their standards; knowledge of various standards is essential. Therefore, ALIM provides extensive expertise on both data integration and data modelling based on linked data and (inter)national open standards [88].

7. BIM-Based Algorithm

The algorithm describes optimization problems with different variables confined by existing conditions and constraints [89]. Solving complex optimization problems in engineering, economy, and science requires utilizing metaheuristic algorithms to find solutions within a reasonable amount of time. Optimization problems are nonlinear, multimodal, and generally subjected to a set of complicated constraints. One of these constraints is the existence of different and conflicting objectives of a single problem which make finding an optimal result difficult [90]. BIM has been considered a comprehensive model for optimizing different construction processes in all stages before and after the project [91]. BIM could provide a suitable framework that supports the decision-making process by utilizing all necessary information at the right time.

Algorithm optimization has been heavily utilized as a decision-making process in construction in the planning of site layout, because such an optimization results in enhancing productivity by facilitating the movement of labor and materials. In addition to saving time and cost, BIM can assist in planning the construction site layout [92]. Implementing this process is not easy, because it relies on many interlocked factors that influence the site layout problem [93]. Hence, it is imperative to create a suitable algorithm to achieve the highest possible case [94]. It has been reported that the optimization process is a complex method, due to the limited number of feasible configurations that lead to obtaining exact methods [95].

The solution seemingly relies on metaheuristic algorithms defined by [96] as a framework of a highly independent algorithm that provides guidelines and strategies for utilizing the optimization process. These algorithms show high compatibility with many engineering optimization problems [97]. Metaheuristic appears in different forms and for various tasks. The main classification of metaheuristic algorithms is either trajectory-based or population-based algorithms. Away from the complications of these algorithms, it can be said that the global utilization of the search algorithms shares the same or very close purposes [98].

7.1. Algorithm Roles in BIM

In recent years, studies have been trying to implement metaheuristic algorithms in BIM-site layouts. As examples, genetic algorithm (GA) and particle swarm optimization (PSO) were among metaheuristic algorithms. The applications targeted the period during planning to support the decision-making process for construction projects [95]. Achieving this algorithm application implies that the site layout is considered in two fashions: a static and a dynamic model. The static model facilitates the initial plans until the end of the entire construction phases [99]. On the other side, dynamic layout models identify the required duration of each facility [100]. The input data and corresponding constraints that characterize the optimal layout problem were discussed by [92]. The optimization process can be mainly achieved by considering the dynamic models, which supposedly contain any possible change throughout the construction phases. In the optimization process, the mathematical procedure may consider integrating generic algorithms to better facilitate the use of a radio frequency identification (RFID) system that depicts tracking object location in the real-time procedure [101]. For the materials, [102] employed a generic algorithm for dynamic plan optimization while using a metaheuristic algorithm to optimize both the material and personal movements. Furthermore, the A* algorithm is used to find the shortest distance between multiple points. Use of the A* algorithm is to counterpart some obstacles in the construction site [103].

Another benefit of using optimization is reducing the time required of travel frequencies at various construction phases [104]. The other application targets other applications in construction projects using integrating BIM product models with several algorithms to achieve optimization, which should be conducted in the auto-generated schedule [105]. As shown later, the BIM simulation system uses a 4D model and generic algorithms to obtain an optimal construction schedule [106]. Furthermore, BIM was utilized to develop and generate construction schedules [107]. These approaches require efficient algorithms to achieve a very high computational speed for optimization using hardware with field-programmable gate arrays [108]. The optimization for maintaining the life cycle cost throughout saving energy was studied by [109].

Sustainability was also studied by integrating the BIM model and the famous multi-objective particle swarm optimization (MOPSO) [110]. BIM is used to reduce the computational time in building fire emergency response operations using the metaheuristic algorithms [108].

7.2. K-Means Algorithm

The tools assessment of the BIM performance focuses on qualitative aspects, such as the success of BIM project progress or the capability of the construction company to complete a project [111]. However, the most challenging part of this type of qualitative assessment is outlining the qualitative BIM performance assessment tools for the operational strategy. Kim et al. [112] have solved this matter by proposing a method based on the k-means clustering algorithm. The solution implies that the BIM performance assessment system enables both the evaluation of the current BIM execution ability and the corresponding prediction of the cost-effectiveness before and after implementation of BIM projects through a comparison with other projects. The method is based on the expectation of a k-means clustering algorithm by analyzing the return-on-investment (RoI) approach.

Previous attempts to solve the same problem relied on various approaches such as implementing BIM initiatives [113], analyzing the BIM substantial benefits and characteristics in estimating the cost-effectiveness [114], analyzing the 13 risk factors (technology, human, management finance, and others) required to consider a swift counterstrategy that results in the success of BIM project or developing six performance indicators of the quality control, schedule conformance, total cost, unit particulars, cost per unit, and safety [115].

Searching for similar project groups can be facilitated using the k-means algorithm, since this algorithm is very efficient in clustering analysis, which quickly results in stable results [112]. A procedure runs the processes associated with the k-means algorithm. The procedure is comprehensive, as it is characterized by input information, metrics to weigh and calculate similarities, and searching for similar projects. The data used for these clusters can be determined by the similarity ratio between the case and target projects, as depicted in Table 3. The metrics are used in evaluating project information and assessing BIM tool, BIM application phase, performance capability, costs, and usage frequency [112].

Table 3. Metrics for analyzing the BIM environment [111].

Metrics	Description
Project information	Project name, type, cost, country, region, etc.
Goal of BIM introduction	Producticvity improvement and schduling reduction in cost.
BIM tool	Revit architecture, Bentley architecture, ArchiCAD, Navisworks, Vico control, etc.
BIM application phase	Design phase-conceptual, construction phase, etc.
Performance capability maturity	Organization, technology, management

7.3. Heuristic Optimisation

A heuristic (a self-discovery) algorithm is a shortcut that allows people to solve problems and make judgments quickly and efficiently, and to find a near-optimal solution. This application means that heuristic algorithms shorten the time to decide, while allowing people to function without constantly stopping to think about their next course of action. This result was subjected to a trade-off balance between several factors, including optimization, accuracy, preciseness, completeness, and solution speed. Heuristic methods were typically employed when the classical solution failed to achieve an exact solution. For large datasets, users must define the objectives to optimize an algorithm using heuristic methods [116]. Despite the potential challenges that visual programming languages (VPLs) pose, they are still offering several benefits. These benefits range from easily creating designing industrial programs with minimal computer science training, to easily accessing APIs platform. Besides, the design is characterized by simple geometry, flexibility, and easy integration that support the automated analysis of non-parametric features [117]. Figure 12 shows the general process for computational algorithm development, which includes various computational procedures, such as Boolean, vectors, and, most importantly, heuristic optimization methods [118]. However, BIM computational processes require many more processes and analyses; hence Seghier et al. [119] proposed different quantitative and qualitative processes which can be computed using algorithms. This type of data analysis forms a specific data management platform that can handle algorithms.

The application of the VLP-algorithm for building design is conventionally practiced according to the workflow presented in Figure 13. Hence, the design can be checked manually by engaging NLP with the auto computational performance. The workflow outlines the initial steps of the BIM model, the logical statement inputs from NLP-based analysis, and the automated compliance checking module.

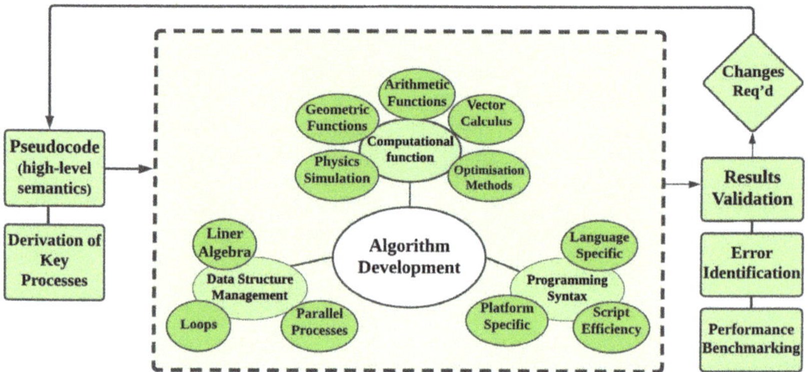

Figure 12. General process for the development of VPL-based computational algorithms for geometric optimization in BIM [112].

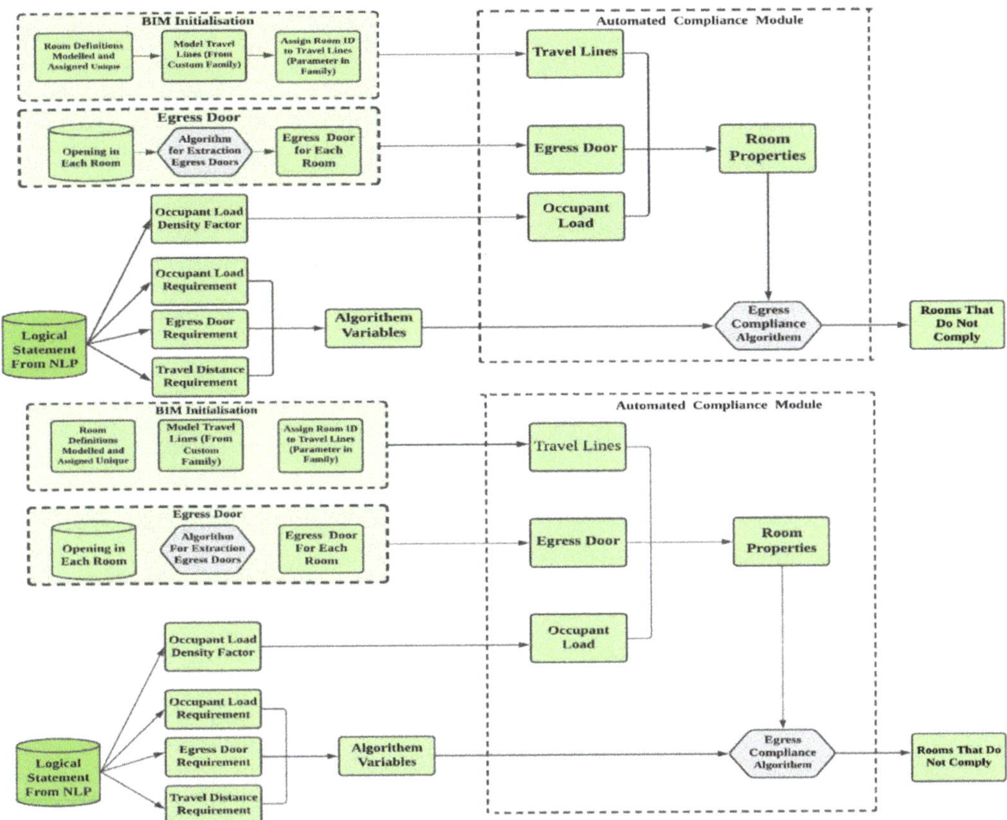

Figure 13. The compliance of the automated and logical building codes using NLP [112].

8. Summary of Previous Empirical Research

Based on the paper title, the role of the interface and interface management, the concept of the multi-model and the relevant applications, and the role of algorithm optimization are highlighted in Table 4. The table shows five elements: objective, methodology, gap(s) presented, and the contribution of each article. One of the most proper methodologies to reduce BIM complexity is the automation, as Mukkavaara [120] reported, who proposed BIM-based automation in the design process. The reason for automation is to manage well in every single application of BIM. This issue was discussed from a different point of view by integrating interface management building with BIM, as comprehensively discussed by Eray et al. [32], aiming at developing a framework that can be used as a good coordinator for communication amongst participants over interface-related problems in the project definition and design phases. The reason for this interface integrating is the ever-increasing complexity of constructing projects, such as power plants and rapid transit systems. Meanwhile, the concept of BIM interface was further discussed by Lin et al. [28] due to the integrating of BIM and web technology, aiming at allowing the user to communicate and explore the various links of BIM.

The main purpose of creating a BIM interface is to allow general contractors to enhance their CMI work efficiency during the construction phase. The BIM interface was introduced by Kang [73] under a different form of human–machine interface (HMI) to monitor the energy management system regarding energy consumption. The results of introducing such as a measure were positive for both the effect and the benefits. Tang et al. [58] have proposed a methodology that targeted smart buildings using a building automation system (BAS). BAS can fill the gap caused by the difficulty of performing the exchange of information amongst BIM stages. This development depends on introducing the network system in IFC by mobilizing the interface. The implementation of BIM faces some barriers in ACE, as reported by Leśniak et al. [121]. They proposed a technique known as the Ishikawa procedure, in which education, including training and studying BIM technology, could result in a better understanding of MIM implementation. The computational BIM is a vast field utilized to automate BIM by optimization to achieve higher efficiency in critical fields, such as better building materials, opening sizes, and glazing types (Lim et al. [122]). The optimization was carried out to serve to integrate Revit tools, dynamo visual programming tools, and multi-objective. As a result, Lim et al. [122] have contributed a series of tools integration using MATLAB to facilitate the possibility of automating and speeding up the process of retrofitting constructions.

The idea of exploring the multimodel approach in BIM was enhanced by Pruvost et al. [6] in projecting uncertainties in designing space by collaborating with the building design workflow. As a result, they integrated several disciplines to share information from different data models and formats, to eventually be used as input in building energy analysis (BEA), including geometry, energy infrastructure, weather, and building usage. Another research study accompanied the progress made by Pruvost et al. [6]. This was in the same field conducted by Fuchs and Scherer [7], who approached BIM multi-model throughout nD-modelling, based on the available original data.

BIM is involved in structural sustainability, as part of the effort to treat the environment better, as reported by Oti et al. [123], who have utilized API in the extended demonstration of the conceptual design option of BIM. This demonstration was conducted by modeling and creating algorithms able to enhance nD building performance. The gap that prompted using nD building was to exploit expanding BIM scope. Another trial to enhance the performance of the multi-model was experienced by Cheng and Cheng [124], in which a genetic algorithm (GA) was employed for better natural selection. The enhancement was carried out by employing Markov chain theory to determine the criteria of adaptive termination with minimal cost. The complexity of MIM applications can be reduced using several techniques.

In 2019, Deng et al. [81] introduced a new parameter in BIM applications as they considered the safety measures a part of the BIM emergency management plan through the Revit platform. In this attempt, Navisworks software was employed. By employing an emergency plan, BIM has become involved in a more detailed approach in construction management. In addition to the work of Pruvost et al. [6] and Lim et al. (2019) in integrating BIM with various techniques as discussed earlier, Li et al. [125] prosed another way to integrate web services with BIM to improve the early design processes called Dynamo BIM, while [126] investigated some approaches to integrate BIM, IoT, and FM for renovating existing buildings. All integration trials mentioned above were validated and then assessed in estimating energy consumption.

The simulation technique is widespread in many fields. In BIM, Siegele et al. [127] used a new MATLAB simulator to study the construction's dynamic feature by programming the object-oriented language of MATLAB. The results have shown that the indoor quality has improved. The last article was chosen to present the optimization needed to upgrade the efficiency of BIM applications in specific fields, such as optimizing the energy consumption and the space occupied by the project.

In this sense, Amiri et al. [94] have suggested a metaheuristic algorithm to support the decision-making process: Uses in planning construction site layout. It has been shown that metaheuristic optimization algorithms have been recognized as very famous hybridizing algorithms that can work very well with the k-means approach.

Table 4. Summary of the most relevant articles.

#	Author/Title	Objective	Methodology	Gap(s)	Contribution
1	Mukkavaara [120]. Structures for supporting BIM-based automation in the design process	To investigate the structures that can be applied to support automation within a BIM-based design process.	Exploring different methods for automation workflows using three studies.	The complexity of the design process s not managed well in each single BIM application.	Providing the foundation for mapping between multiple sets of data to resolve the coupling of information at each activity in an automated BIM-based workflow.
2	Eray et al. [32]. An overview on integrating interface management and building information management systems	To develop a framework for integrating IM and BIM systems to create better coordination and communication between project participants over interface related problems in the project definition and design phases.	By explaining the relation between IM and BIM systems, a framework was developed to connect interface points within a 3-D model followed by validating the functionality of proposed framework.	The complex projects such rapid transit systems, power plants, refineries and port facilities were facing difficulties in managing execution due to geographical specialized location.	Integrating IM and BIM systems improves better execution process by improving the project control, communications and alignment along with reduced requests for information, change requests, and rework.
3	Tang et al. [58]. BIM assisted Building Automation System information exchange using BACnet and IFC	To link smart buildings to different building systems together with the Building Automation System (BAS).	Use IDM and MVD methodologies to define an IFC subset schema so that BAS information conforming to the BACnet protocol can be represented in IFC data model for information exchange throughout various project stages with BIM tools.	It is rarely seen to design BAS or exchange BAS information in different project stages using BIM tools.	Facilitate information exchange for BIM-assisted BAS design and operation using one BAS open communication protocol named Building Automation and Control Networks (BACnet) and open BIM standard Industry Foundation Class (IFC).
4	Lin et al. [28]. Construction database-supported and BIM-based interface communication and management: a pilot project	To integrate BIM and web technology to construct projects allows users to communicate interface issues and obtain responses for them effectively.	Using a case (pilot) study in a building project by proposing database communication and management interface (dBCMI) system.	The absence of suitable and necessary systems or platforms to tackle the communication and interface management.	Developing a database-supported and BIM-based CMI (DBCMI) system for general contractors to enhance their CMI work efficiency during the construction phase.
5	Kang [73]. BIM-based human–machine interface (HMI) framework for energy management	To introduce Building Information Modeling (BIM)-based Human–Machine Interface (HMI) framework for intuitive space-based energy management.	Introducing BIM-based HMI framework after deriving the considerations and requirements necessary for linking the energy control point and BIM through a questionnaire designed by practitioners.	The absence of effective heterogeneous link between BIM and energy management system to provides space-based real-time energy monitoring.	A positive effect (3.9/5.0) on the connectivity of BIM-based HMI with benefits (4.3/5.0) for real-time data monitoring.
6	Leśniak et al. [121]. Barriers to BIM Implementation in Architecture, Construction, and Engineering Projects—The Polish Study	To analyze the cause and effect of identified barriers (failure) to implementing BIM technology in the construction process in Poland.	Employing a tool that helps to recognize the actual or potential causes of failure known as Ishikawa.	Limited information about the influence of the poor BIM implementation in Poland and about the awareness of reducing the obstacles of BIM implementation.	Introducing factors that are needed to better implement BIM, such as education, training, and studying BIM technology
7	Lim et al. [122]. Computational BIM for Green Retrofitting of The Existing Building Envelope	To automate the computational building information modelling (BIM) in decision-making for green retrofitting of the existing building.	Integrating Revit tool, dynamo visual programming tool, and multi-objective optimization algorithm to optimize overall thermal transfer value (OTTV) and construction investment cost.	The need of better and efficient decision to optimize the building efficiency, such as the choices of building materials, opening sizes, and glazing types.	The integration (Revit), VPL (Dynamo), and MOO (NSGA-II in MATLAB) facilitates the possibility of automating and speeding up the process of green retrofitting performance.

Table 4. *Cont.*

#	Author/Title	Objective	Methodology	Gap(s)	Contribution
8	Pruvost et al. [6]. Multimodel-based exploration of the building design space and its uncertainty	To support analysis of uncertainty by presenting an innovative modeling approach that collaborates building design workflow.	Integrating several disciplines to share information from different data models and formats to eventually be used as input in building energy analysis (BEA) including geometry, energy infrastructure, weather, building usage.	Computational methods lack rapid and more mature methods for designing options to find the best alternative.	The multimodal method is extended for a broad exploration of building design options and their inherent uncertainty.
9	Oti et al. [123]. Structural sustainability appraisal in BIM	To utilize API in BIM extension and demonstrates its application to embed sustainability issues in the structural conceptual design options in BIM.	The approach was achieved by mapping API for structural sustainability appraisal followed by developing assessment model and integrating this model using conceptual building design iterations.	APIs are not yet fully exploited in expanding the BIM scope.	The utilization process has expanded the BIM scope by modelling and creating algorithms applicable to enhance nD building performance.
10	Fuchs & Scherer [7]. Multimodels—Instant nD-modeling using original data	To introduce multimodels approach to offer wider value of information in terms of quality and time.	The loose cross-model coupling of data elements is neutrally stored in external ID-based link models.	The current interoperability lacks generality and satisfaction.	Offering multimodel approach to the single self-contained information space.
11	Cheng & Cheng [124]. Enhancing Multi-model Inference with Natural Selection	To employ the genetic algorithm (GA) to inspire the process of natural selection using crossover and mutation iteratively to update a collection of potential solutions (models) until convergence.	The use of the Markov chain theory to design an adaptive termination criterion that vastly reduces the computational cost.	The studies on the availability of candidate qualified models are very rare in literature.	It developed a new schema theory that characterizes the current model to improve the evolutionary process by demonstrating the GA empirical power based on two real data examples.
12	Deng et al. [81]. Research on safety management application of dangerous sources in engineering construction based on BIM technology	To create a construction hazard source safety management module through secondary development of the Revit platform.	Using the simulator Navisworks software to rescue the emergency of construction safety accidents by formulating corresponding emergency management plan	Existing hidden accidental dangers in construction without proper solution.	Introducing the security management module to guide developers to avoid accidents.
13	Li et al. [125]. Integration of Building Information Modeling and Web Service Application Programming Interface for assessing building surroundings in early design stages	To integrate Dynamo BIM and Amap web service APIs for the evaluations of diverse uses of transportations.	Results from the integrated tool are analyzed and validated with survey results	Developing service tools relates BIM and location to facilitate the process of estimating energy consumption.	The integration of Dynamo BIM and web service APIs is helpful for site assessments in the early design stage or even earlier.
14	Siegele et al. [127]. A new MATLAB Simulink Toolbox for Dynamic Building Simulation with BIM and Hardware in the Loop compatibility	To develop carnotUIBK for dynamic building simulations using MATLAB technique.	The development of this tool is carried out using programming the object-oriented language of MATLAB.	Research work using simulator MATLAB is very rare in BIM dynamic and loop compatibility.	The new model is tailored for hardware in the loop applications development and indoor air quality simulations in terms of multi-zone modelling.
15	Amiri et al. [95]. BIM-based applications of metaheuristic algorithms to support the decision-making process: Uses in the planning of construction site layout	To introduce BIM-based applications of metaheuristic optimization algorithms to support the decision-making process.	Metaheuristic optimization algorithms was employed by hybridizing several algorithms such as k-means approach.	Optimization has been treated inefficiently in most off the previous work especially in site layout planning.	BIM has been equipped with the optimized decision-making process by aggregating the necessary information at the right time.

9. Contribution

Adopting algorithms in interface and IM in BIM multi-model applications for optimization has been considered the most significant contributor to achieving high performance in construction. This novel contribution was based on introducing suitable methodologies to accomplish new or upcoming research tasks. The process cannot be accomplished without specifying API in various software components that may interact, such as accessing the database, hard drive, disc drive, video card, etc. The interface is established to create programming codes equipped with programming language routines, data structures, and classes and variables. This review explains the integration of IBS and MBS in BIM and interfaces management. The other contribution of this study is to develop an API to enable BIM viewers to simplify BIM-based interface management.

One of the significant advanced steps to extend BIM use was developing a BIM-IoT system that allows one to directly use the BIM models for building context and 3D views. Alternatively, considering the BIM Model management in a native environment, VPL scripts have been developed to support the integration of BIM and IoT, which requires employing sensors.

VPL can also help introduce an accessible mode that enables BIM/IoT interfacing, in which BIM models could be transferred from static to dynamic, with the ability to self-update essential information. BIM/IoT opens a new trend of communication called machine-to-machine (M2M) communication, which extends to using external databases between the real and virtual worlds.

The review also presented an approach for a workflow engine to be integrated into the BIM multi-model collaboration platform. This type of implementation is considered the most critical step towards increasing the degree of automation. Another field, in addition to automation, was to create a multi-model BIM collaboration platform based on ontologies and BCF. The review has shown that the workflow of specific data can be obtained from the multi-model IDM/MVD methodology to achieve process and user model and their linkage information.

The review has introduced factors needed to implement better BIM, such as education, training, and studying BIM technology. In addition, BIM has expanded in scope as the algorithms were introduced to enhance nD building performance. Utilizing algorithms in BIM processes improved the process by demonstrating the GA empirical power based on two real data examples. Furthermore, integrating Dynamo BIM and web service APIs could improve site assessments in the early design stage, or even earlier.

10. Conclusions and Future Work

The review has shown clearly that there is still a long way to reach the objective of high performance of BIM. The interoperability issues were identified and reviewed; however, they need to be addressed in future versions of BIM tools schemas. The fundamental problem of these issues is slowing down the BIM automation and in areas that still employ manual interaction. The currently available tools likely can read, with minimum error, the BIM files. Most issues occur either when the BIM files are generated by CAD tools such as Revit or, more importantly, during reading the BIM files by energy simulation tools—a field that needs more focus and understanding in future research areas. The homogeneous data format can be easily used in the multi-model strategies instead of the generic data access to the elementary models, which requires creating a virtual structure throughout generalizing and idealizing the representation of popular data format concepts. Under this generalization, BIM applications can be extended without being conceptually changed, or maintaining a super data model.

The review has identified the possibility of utilizing the BACnet protocol for BAC information exchange that considers the IFC data model. Hence, the new technique can unlock the potential future of smart buildings using BIM and a BIM multi-model approach, which can be achieved by integrating tools and with BIM through IoT.

Despite the advancement of BIM in construction, the need for optimization BIM processes still represents the core of current and future developments. Recently, researchers have been working on adopting metaheuristic algorithms to support making decisions in construction by considering BIM-based applications. These applications include site layout to identify the size, energy consumption, and possible constraints concerning optimum cost outcomes. The algorithm's involvement has successfully enhanced productivity and safety in the construction process, saving cost and time by creating an intelligent system to control moving labor and materials. The optimization has shown clear evidence of the effectiveness of aggregating the necessary information at the right time.

The review presents limitations, especially in considering the potential link of BIM and Building Management System (BMS), and the level of influencing the BIM-IoT prototype. It was also revealed that the solution proposed by BIM skills to solve FM management in the dynamic model is still an unusual scenario because of the limited contribution of the BIM model in associating BMS environment.

The applications of interface management have been reviewed academically and practically in the literature, suggesting numerous BIM-based system developments in which CMI work is based on the filing system. This filing system has shortcomings which, alternatively, are replaced by proposing a DBCMI system, due to its capability to overcome the limitations in the other filing system. Employing a DBCMI system relies on establishing more effective visualization and sharing for BIM-based interface management during the project construction phase. It effectively helps integrate discourse in the BIM model and to improve the communication of information. Furthermore, designing API modules to be used in the DBCMI system simplifies using interface and operations that increase the willingness of participants to use the system. Despite the advancement of BIM in construction, the need to optimize BIM processes still represents the core of current and future developments.

The review presented the need for optimization by extending the contribution of the various algorithms in this process. Recently, researchers have been working on adopting metaheuristic algorithms to support deciding on construction by considering BIM-based applications. These applications include site layout to identify the size, energy consumption, and possible constraints concerning optimum cost outcomes. The algorithm's involvement has successfully shown an enhancement in productivity and safety in the construction process, which saves cost, time, and the framework of moving labor and materials. The optimization has shown clear evidence of the effectiveness of aggregating the necessary information at the right time. It was also revealed that the solution proposed by BIM skills to solve FM management in the dynamic model is still an unusual scenario because of the limitation of the BIM model contribution that is associated with the BMS environment.

Author Contributions: Conceptualization, N.A.H. (Nawal Abdunasseer Hmidah) and A.B.A.A.; methodology, N.A.H. (Nuzul Azam Haron); writing—original draft preparation, N.A.H. (Nawal Abdunasseer Hmidah), writing—review and editing, N.A.H. (Nuzul Azam Haron) supervision and reviewed the article, N.A.H. (Nuzul Azam Haron) advice and support, A.H.A. and T.H.L. and R.A.A.R.A.E. All authors have read and agreed to the published version of the manuscript.

Funding: This research received no external funding. This work was supported and partial funding by research management center, University Putra Malaysia.

Institutional Review Board Statement: Not applicable.

Informed Consent Statement: Not applicable.

Data Availability Statement: Not applicable.

Acknowledgments: The authors would like to thank the Department of Civil Engineering, Faculty of Engineering, University Putra Malaysia, Serdang, Malaysia, for supporting This review paper.

Conflicts of Interest: The authors declare no conflict of interest.

References

1. Baiardi, L.; Ferreira, E.A.M. The integrated project for the redevelopment of a historic building. In *Handbook of Research on Industrial Informatics and Manufacturing Intelligence*; IGI Global: Hershey, PA, USA, 2020; pp. 261–282.
2. Ahuja, R.; Sawhney, A.; Arif, M. Driving lean and green project outcomes using BIM: A qualitative comparative analysis. *Int. J. Sustain. Built Environ.* **2017**, *6*, 69–80. [CrossRef]
3. Gledson, B.J.; Greenwood, D. The adoption of 4D BIM in the UK construction industry: An innovation diffusion approach. *Eng. Constr. Arch. Manag.* **2017**, *24*, 950–967. [CrossRef]
4. Yang, A.; Han, M.; Zeng, Q.; Sun, Y. Adopting building information modeling (BIM) for the development of smart buildings: A review of enabling applications and challenges. *Adv. Civ. Eng.* **2021**, *2021*, 8811476. [CrossRef]
5. Ivson, P.; Moreira, A.; Queiroz, F.; Santos, W.; Celes, W. A systematic review of visualization in building information modeling. *IEEE Trans. Vis. Comput. Graph.* **2020**, *26*, 3109–3127. [CrossRef] [PubMed]
6. Pruvost, H.; Katranuschkov, P.; Scherer, R. Multimodel-based exploration of the building design space and its uncertainty. *Proceedings* **2017**, *1*, 1078. [CrossRef]
7. Fuchs, S.; Scherer, R.J. Multimodels—Instant nD-modeling using original data. *Autom. Constr.* **2017**, *75*, 22–32. [CrossRef]
8. Honti, R.; Erdélyi, J. Possibilities of bim data exchange. In Proceedings of the 18th International Multidisciplinary Scientific GeoConference SGEM, Albena, Bulgaria, 2–8 July 2018; Volume 18, pp. 923–930. [CrossRef]
9. Eastman, C.; Teicholz, P.; Sacks, R.; Liston, K. *BIM Handbook: A Guide to Building Information Modeling for Owners, Managers, Designers, Engineers and Contractors*; John Wiley and Sons: New York, NY, USA, 2008.
10. Schwabe, K.; Teizer, J.; König, M. Applying rule-based model-checking to construction site layout planning tasks. *Autom. Constr.* **2019**, *97*, 205–219. [CrossRef]
11. Xiang, W.; Wang, F.; Zhang, X. Research on Integration of multi-source BIM models based on gis platform. In Proceedings of the 2019 5th International Conference on Information Management (ICIM 2019), Cambridge, UK, 24–27 March 2019; pp. 40–44. [CrossRef]
12. Heaton, J.; Parlikad, A.K.; Schooling, J. Design and development of BIM models to support operations and maintenance. *Comput. Ind.* **2019**, *111*, 172–186. [CrossRef]
13. Kreider, R.; Messner, J.; Dubler, C. Determining the frequency and impact of applying BIM for different purposes on projects. In Proceedings of the 6th International Conference on Innovation in Architecture, Engineering and Construction (AEC), Philadelphia, PA, USA, 9–11 June 2010; pp. 9–11.
14. Sacks, R.; Eastman, C.; Teicholz, P.; Lee, G. *Manual de BIM-: Um Guia de Modelagem da Informação da Construção Para Arquitetos, Engenheiros, Gerentes, Construtores e Incorporadores*; Bookman Editora: Stockholm, Sweden, 2014.
15. Ramaji, I.J.; Messner, J.I.; Leicht, R.M. Leveraging building information models in IFC to perform energy analysis in openstudio®. In Proceedings of the ASHRAE and IBPSA-USA simBuild 2016 Building Performance Modeling Conference, Salt Lake City, UT, USA, 8–12 August 2016; pp. 251–258.
16. Chong, H.-Y.; Lee, C.-Y.; Wang, X. A mixed review of the adoption of building information modelling (BIM) for sustainability. *J. Clean. Prod.* **2017**, *142*, 4114–4126. [CrossRef]
17. Khodeir, L.M.; Nessim, A.A. BIM2BEM integrated approach: Examining status of the adoption of building information modelling and building energy models in Egyptian architectural firms. *Ain Shams Eng. J.* **2018**, *9*, 1781–1790. [CrossRef]
18. Samuel, E.I.; Joseph-Akwara, E.; Richard, A. Assessment of energy utilization and leakages in buildings with building information model energy. *Front. Arch. Res.* **2017**, *6*, 29–41. [CrossRef]
19. Tu, K.J.; Vernatha, D. Application of Building Information Modeling in energy management of individual departments occupying university facilities. *Int. J. Archit. Environ. Eng.* **2016**, *10*, 225–231.
20. Alahmad, M.; Nader, W.; Neal, J.; Shi, J.; Berryman, C.; Cho, Y.; Lau, S.-K.; Li, H.; Schwer, A.; Shen, Z.; et al. Real Time Power Monitoring & integration with BIM. In Proceedings of the IECON 2010-36th Annual Conference on IEEE Industrial Electronics Society, Glendale, AZ, USA, 7–10 November 2010; pp. 2454–2458.
21. Woo, J.H.; Diggelman, C.; Abushakra, B. BIM-based energy monitoring with XML parsing engine. In Proceedings of the 28th ISARC, Seoul, Korea, 29 June 2011; pp. 544–545.
22. Jrade, A.; Jalaei, F. Integrating building information modelling (BIM) and energy analysis tools with green building certification system to conceptually design sustainable buildings. *J. Inf. Technol. Constr.* **2014**, *19*, 494–519.
23. Wren, D.A. Interface and interorganizational coordination. *Acad. Manag. J.* **1967**, *10*, 69–81. [CrossRef]
24. Lin, Y.-C. Developing Construction Network-Based Interface Management System. In Proceedings of the Construction Research Congress 2009, Seattle, WA, USA, 5–7 April 2009; pp. 477–486.
25. Shokri, S. Interface Management for Complex Capital Projects. Ph.D. Thesis, University of Waterloo, Waterloo, ON, Canada, 2014.
26. Shen, W.; Tang, W.; Wang, S.; Duffield, C.F.; Hui, F.K.P.; You, R. Enhancing trust-based interface management in international engineering-procurement-construction projects. *J. Constr. Eng. Manag.* **2017**, *143*, 04017061. [CrossRef]
27. Eray, E.; Sanchez, B.; Haas, C. Usage of interface management system in adaptive reuse of buildings. *Buildings* **2019**, *9*, 105. [CrossRef]
28. Lin, Y.-C.; Jung, S.; Su, Y.-C. Construction database-supported and BIM-based interface communication and management: A pilot project. *Adv. Civ. Eng.* **2019**, *2019*, 8367131. [CrossRef]

29. Pavitt, T.C.; Gibb, A.G.F. Interface management within construction: In particular, building facade. *J. Constr. Eng. Manag.* **2003**, *129*, 8–15. [CrossRef]
30. Critsinelis, A. The modern field development approach. In Proceedings of the OMAE 20th International Conference on Offshore Mechanics and Arctic Engineering, Rio de Janeiro, Brazil, 3–8 June 2001; Volume 1, pp. 391–400.
31. Laan, J.; Wildenburg, L.; Kleunen, P. Dynamic interface management in transport infrastructure project. In Proceedings of the 2nd European Systems Engineering Conference, Munich, Germany, 13–15 September 2000; pp. 13–15.
32. Eray, E.; Golzarpoo, B.; Rayside, D.; Haas, C. An overview on integrating interface management and building information management systems. In Proceedings of the 6th CSCE-CRC International Construction Specialty Conference 2017—Held as Part Canadian Society Civil Engineering Annual Conference and General Meeting, Vancouver, BC, Canada, 31 May–3 June 2017; Volume 1, pp. 196–205.
33. Lin, Y.-C. Use of BIM approach to enhance construction interface management: A case study. *J. Civ. Eng. Manag.* **2015**, *21*, 201–217. [CrossRef]
34. Harris, F.; McCaffer, R.; Baldwin, A.; Edum-Fotwe, F. *Modern Construction Management*; John Wiley & Sons: Hoboken, NJ, USA, 2021.
35. Chen, Q.; Reichard, G.; Beliveau, Y. An interface object model for interface management in building construction. In *Proceedings of the eWork and eBusiness in Architecture, Engineering and Construction*; CRC Press: Boca Raton, FL, USA, 2020; pp. 119–126.
36. Ma, Z.; Ren, Y.; Xiang, X.; Turk, Z. Data-driven decision-making for equipment maintenance. *Autom. Constr.* **2020**, *112*, 103103. [CrossRef]
37. Chen, Q.; Reichard, G.; Beliveau, Y. Multiperspective approach to exploring comprehensive cause factors for interface issues. *J. Constr. Eng. Manag.* **2008**, *134*, 432–441. [CrossRef]
38. Morris, P.W.G. Managing project interfaces—Key points for project success. In *Project Management Handbook*; Hoboken, NJ, USA, 2008; pp. 16–55. [CrossRef]
39. Stuckenbruck, L.C. Integration: The essential function of project management. In *Project Management Handbook*; Hoboken, NJ, USA, 2008; pp. 56–81. [CrossRef]
40. Caglar, J.P.E.; Connolly, M. Interface Management: Effective Information Exchange through Improved Communication. ABB Value Paper Series. 2007. Available online: https://docplayer.net/10602195-Interface-management-effective-information-exchange-through-improved-communication-abb-value-paper-series.html (accessed on 1 September 2021).
41. Chen, Q.; Reichard, G.; Beliveau, Y. Interface management—A facilitator of lean construction and agile project management. In Proceedings of the Proceedings IGLC-15, East Lansing, MI, USA, 15 July 2007; pp. 57–66.
42. Park, J.; Cai, H.; Dunston, P.S.; Ghasemkhani, H. Database-supported and web-based visualization for daily 4D BIM. *J. Constr. Eng. Manag.* **2017**, *143*, 04017078. [CrossRef]
43. Shokri, S.; Haas, C.T.; Haas, R.C.G.; Lee, S. Interface-management process for managing risks in complex capital projects. *J. Constr. Eng. Manag.* **2016**, *142*, 04015069. [CrossRef]
44. Cotton, I.W.; Licklider, F.S.G. Data structures and techniques for remote computer graphics. In Proceedings of the Fall Joint Computer Conference, San Francisco, CA, USA, 9 December 1968; Volume 33, pp. 533–544. [CrossRef]
45. De, B. API management. In *API Management*; Apress: Berkeley, CA, USA, 2017; pp. 15–28. [CrossRef]
46. Tan, W.; Fan, Y.; Ghoneim, A.; Hossain, M.A.; Dustdar, S. From the service-oriented architecture to the web api economy. *IEEE Internet Comput.* **2016**, *20*, 64–68. [CrossRef]
47. Robillard, M.P. What makes apis hard to learn? Answers from developers. *IEEE Softw.* **2009**, *26*, 27–34. [CrossRef]
48. Hou, L.; Zhao, S.; Li, X.; Chatzimisios, P.; Zheng, K. Design and implementation of application programming interface for Internet of things cloud. *Int. J. Netw. Manag.* **2017**, *27*, e1936. [CrossRef]
49. Shnier, M. *Dictionary of PC Hardware and Data Communications Terms*; Society for Technical Communication: Sebastopol, CA, USA, 1996.
50. Koch, F.J. *Opportunities and Barriers for Advancing the API Economy within the Automotive Industry*; Technische Universität München: Munich, Germany, 1996.
51. Kansa, E.C.; Wilde, E. Tourism, peer production, and location-based service design. In Proceedings of the 2008 IEEE International Conference on Services Computing, Honolulu, HI, USA, 7–11 July 2008; Volume 2, pp. 629–636.
52. Qiu, Y. The openness of open application programming interfaces. *Inf. Commun. Soc.* **2016**, *20*, 1720–1736. [CrossRef]
53. Santoro, M.; Vaccari, L.; Mavridis, D.; Smith, R.; Posada, M.; Gattwinkel, D. *Web Application Programming Interfaces (Apis): General Purpose Standards, Terms and European Commission Initiatives*; Publications Office of the European Union: Luxembourg, 2019.
54. Sheng Jun, T.; Qing, Z.; Junqiao, Z. Towards interoperating of BIM and GIS model: Geometric and semantic integration of CityGML and IFC building models. *J. Inf. Technol. Civ. Eng. Archit.* **2014**, *6*, 11–17.
55. Zhang, J. Towards systematic understanding of geometric representations in bim standard: An empirical data-driven approach. In Proceedings of the Construction Research Congress 2018: Construction Information Technology, New Orleans, LA, USA, 2–4 April 2018; pp. 96–105. [CrossRef]
56. Afsari, K.; Eastman, C.M.; Castro-Lacouture, D. JavaScript Object Notation (JSON) data serialization for IFC schema in web-based BIM data exchange. *Autom. Constr.* **2017**, *77*, 24–51. [CrossRef]
57. Alshehri, F.; Kenny, P.; Pinheiro, S.V.; Ali, U.; O'Donnell, J. Development of a Model View Definition (MVD) for ther-mal comfort analysis in commercial buildings using BIM and EnergyPlus. In Proceedings of the CitA BIM Gather 2017, Croke Park, Dublin, Ireland, 23–24 November 2017.

58. Tang, S.; Shelden, D.R.; Eastman, C.M.; Pishdad-Bozorgi, P.; Gao, X. BIM assisted building automation system information exchange using BACnet and IFC. *Autom. Constr.* **2020**, *110*, 103049. [CrossRef]
59. Sattler, L.; Lamouri, S.; Pellerin, R.; Maigne, T. Interoperability aims in building information modeling exchanges: A literature review. *IFAC PapersOnLine* **2019**, *52*, 271–276. [CrossRef]
60. Hon, C.K.; Sun, C.; Xia, B.; Jimmieson, N.L.; Way, K.A.; Wu, P.P.-Y. Applications of Bayesian approaches in construction management research: A systematic review. *Eng. Constr. Arch. Manag.* **2021**. Available online: https://www.emerald.com/insight/content/doi/10.1108/ECAM-10-2020-0817/full/html (accessed on 1 September 2021). [CrossRef]
61. Lee, Y.-C.; Eastman, C.M.; Solihin, W. Rules and validation processes for interoperable BIM data exchange. *J. Comput. Des. Eng.* **2021**, *8*, 97–114. [CrossRef]
62. Lee, Y.-C.; Eastman, C.M.; Solihin, W. Logic for ensuring the data exchange integrity of building information models. *Autom. Constr.* **2018**, *85*, 249–262. [CrossRef]
63. Stanojević, V.; Vlajic, S.; Milic, M.; Ognjanović, M.; Vlajić, S.A.; Milić, M. Guidelines for framework development process. In Proceedings of the 2011 7th Central and Eastern European Software Engineering Conference (CEE-SECR), Moscow, Russia, 31 October–3 November 2011. [CrossRef]
64. Hanington, B.; Martin, B. *Universal Methods of Design Expanded and Revised: 125 Ways to Research Complex. Problems, Develop Innovative Ideas, and Design Effective Solutions*; Rockport Publishers: Beverly, MA, USA, 2019.
65. Van Deursen, A.; Klint, P.; Visser, J. Domain-specific languages. *ACM SIGPLAN Not.* **2000**, *35*, 26–36. [CrossRef]
66. Mattsson, M.; Bosch, J. Characterizing stability in evolving frameworks. In Proceedings of the Proceedings Technology of Object-Oriented Languages and Systems. TOOLS 29 (Cat. No.PR00275), Nancy, France, 7–10 June 1999; pp. 118–130.
67. Snaider, J.; McCall, R.; Franklin, S. The LIDA framework as a general tool for AGI. In *Lecture Notes Computer Science*; Springer: Berlin/Heidelberg, Germany, 2011; Volume 6830. [CrossRef]
68. Pan, Y.; Zhang, L. A BIM-data mining integrated digital twin framework for advanced project management. *Autom. Constr.* **2021**, *124*, 103564. [CrossRef]
69. Wu, C.-S.; Simmons, D.B. Software Project Planning Associate (SPPA): A knowledge-based approach for dynamic software project planning and tracking. In Proceedings of the Proceedings 24th Annual International Computer Software and Applications Conference, COMPSAC2000, Taipei, Taiwan, 25–27 October 2002; pp. 305–310. [CrossRef]
70. Zhang, H.; Kitchenham, B.; Jeffery, R. Planning Software Project Success with Semi-Quantitative Reasoning. In Proceedings of the 19th Australian Conference on Software Engineering (ASWEC), Melbourne, VIC, Australia, 10–13 April 2007; pp. 369–378. [CrossRef]
71. Grant, C.; Osanloo, A. University understanding, selecting, and integrating a theoretical framework in dissertation research: Creating the blueprint for your "house". *Adm. Issues J. Educ. Prac. Res.* **2014**, *4*, 12–26. [CrossRef]
72. Alnaggar, A.; Pitt, M. Towards a conceptual framework to manage BIM/COBie asset data using a standard project management methodology. *J. Facil. Manag.* **2019**, *17*, 175–187. [CrossRef]
73. Kang, T. BIM-based human machine interface (hmi) framework for energy management. *Sustainability* **2020**, *12*, 8861. [CrossRef]
74. Siao, F.-C.; Shu, Y.-C.; Lin, Y.-C. Interface management practices in Taiwan construction project. In Proceedings of the 28th International Symposium on Automation and Robotics in Construction (ISARC), Taipei, Taiwan, 28 June–1 July2011; pp. 947–952. [CrossRef]
75. Kerzner, H. *Project Management Project*; Jone and Wiley and Sons: Hoboken, NJ, USA, 2009.
76. Evans, G.; Bailey, A.; Reed, M.; Naim, M.; Brown, D.; Towill, D.; Riley, M. Organising for improved construction interfaces. *Innov. Civ. Constr. Eng.* **2009**, *49*, 207–212. [CrossRef]
77. Gould, F.; Joyce, N. *Construction Project Management*, 4th ed.; Pearson: London, UK, 2013.
78. Ding, L.; Zhou, Y.; Luo, H.; Wu, X. Using nD technology to develop an integrated construction management system for city rail transit construction. *Autom. Constr.* **2012**, *21*, 64–73. [CrossRef]
79. Vainiunas, P.; Popovas, V.; Jarmolajev, A. Non-linear 3D modelling of RC slab punching shear failure. *J. Civ. Eng. Manag.* **2004**, *10*, 311–316. [CrossRef]
80. Vanlande, R.; Nicolle, C.; Cruz, C. IFC and building lifecycle management. *Autom. Constr.* **2008**, *18*, 70–78. [CrossRef]
81. Deng, L.; Zhong, M.; Liao, L.; Peng, L.; Lai, S. Research on safety management application of dangerous sources in engineering construction based on BIM technology. *Adv. Civ. Eng.* **2019**, *2019*, 7450426. [CrossRef]
82. Gorzalka, P.; Schmiedt, J.E.; Schorn, C.; Hoffschmidt, B. Automated generation of an energy simulation model for an existing building from UAV imagery. *Buildings* **2021**, *11*, 380. [CrossRef]
83. Cheng, M.Y.; Chiu, K.C.; Hsieh, Y.M.; Yang, I.T.; Chou, J.S.; Wu, Y.W. BIM integrated smart monitoring tech-nique for building fire prevention and disaster relief. *Autom. Constr.* **2017**, *84*, 14–30. [CrossRef]
84. Lin, Y.-C. Construction network-based interface management system. *Autom. Constr.* **2013**, *30*, 228–241. [CrossRef]
85. Ju, Q.; Ding, L.; Skibniewski, M.J. Optimization strategies to eliminate interface conflicts in complex supply chains of construction projects. *Vilnius Gedim. Tech. Univ.* **2017**, *23*, 712–726. [CrossRef]
86. Eray, E. Measuring and Visualizing Integrated Project Interface Status. Ph.D. Thesis, University of Waterloo, Waterloo, ON, Canada, 2020.
87. Hull, J.; Ewart, I.J. Conservation data parameters for BIM-enabled heritage asset management. *Autom. Constr.* **2020**, *119*, 103333. [CrossRef]

88. Wang, W.; Hu, H.; Zhang, J.; Hu, Z. Digital twin-based framework for green building maintenance system. In Proceedings of the 2020 IEEE International Conference on Industrial Engineering and Engineering Management (IEEM), Singapore, 14–17 December 2020; pp. 1301–1305.
89. Diakite, A.A.; Zlatanova, S. Automatic geo-referencing of BIM in GIS environments using building footprints. *Comput. Environ. Urban. Syst.* **2020**, *80*, 101453. [CrossRef]
90. Suroliya, M.; Dhaka, V.S.; Poonia, R.C. Metaheuristics for communication protocols: Overview and conceptual comparison. *Int. J. Adv. Res. Comput. Sci. Softw. Eng.* **2014**, *4*, 353–360.
91. Salihi, I.U. Application of structural building information modeling (S-BIM) for sustainable buildings design and waste reduction: A review. *Int. J. Appl. Eng. Res.* **2016**, *11*, 1523–1532.
92. Krepp, S.; Jahr, K.; Bigontina, S.; Bügler, M.; Borrmann, A. BIMsite—Towards a BIM-based generation and evaluation of realization variants comprising construction methods, site layouts and schedules. In Proceedings of the 23rd International Workshop on Intelligent Computing in Engineering (EG-ICE 2016), Krakow, Polish, 29 June–1 July 2016; pp. 1–10.
93. Astour, H.; Franz, V. BIM-and simulation-based site layout planning. In Proceedings of the 2014 International Conference on Computing in Civil and Building Engineering, Orlando, FL, USA, 23–25 June 2014; pp. 291–298. [CrossRef]
94. Chahrour, R. *Integration von CAD und Simulation auf Basis von Produktmodellen im Erdbau*; Kassel University Press GmbH: Kassel, Germany, 2007.
95. Amiri, R.; Sardroud, J.M.; De Soto, B.G. BIM-based applications of metaheuristic algorithms to support the decision-making process: Uses in the planning of construction site layout. *Procedia Eng.* **2017**, *196*, 558–564. [CrossRef]
96. Sörensen, K.; Glover, F. Metaheuristics. Encyclopedia of Operations Research and Management Science. In *Encyclopedia of Operations Research and Management Science*; Gass, S.I., Harris, C.M., Eds.; Springer: New York, NY, USA, 2013; pp. 960–970.
97. Alba, E.; García-Nieto, J.; Taheri, J.; Zomaya, A. New research in nature inspired algorithms for mobility management in gsm networks. In *Lecture Notes in Computer Science*; Springer: Berlin/Heidelberg, Germany, 2008; Volume 4974. [CrossRef]
98. Yang, X.-S. Metaheuristic optimization. *Scholarpedia* **2011**, *6*, 11472. [CrossRef]
99. Osman, H.M.; Georgy, M.E.; Ibrahim, M.E. A hybrid CAD-based construction site layout planning system using genetic algorithms. *Autom. Constr.* **2003**, *12*, 749–764. [CrossRef]
100. Yahya, M.; Saka, M.P. Construction site layout planning using multi-objective artificial bee colony algorithm with Levy flights. *Autom. Constr.* **2014**, *38*, 14–29. [CrossRef]
101. Akanmu, A.; Olatunji, O.; Love, P.E.D.; Nguyen, D.; Matthews, J. Auto-generated site layout: An integrated approach to real-time sensing of temporary facilities in infrastructure projects*. *Struct. Infrastruct. Eng.* **2015**, *12*, 1243–1255. [CrossRef]
102. Yu, Q.; Li, K.; Luo, H. A BIM-based dynamic model for site material supply. *Procedia Eng.* **2016**, *164*, 526–533. [CrossRef]
103. Kumar, S.S.; Cheng, J.C.P. A BIM-based automated site layout planning framework for congested construction sites. *Autom. Constr.* **2015**, *59*, 24–37. [CrossRef]
104. Hammad, A.W.A.; Akbarnezhad, A.; Rey, D.; Waller, S.T. A computational method for estimating travel frequencies in site layout planning. *J. Constr. Eng. Manag.* **2016**, *142*, 04015102. [CrossRef]
105. Liu, S. Sustainable building design optimization using building information modeling. In Proceedings of the ICCREM 2015, Luleå, Sweden, 11–12 August 2015; pp. 326–335. [CrossRef]
106. Moon, H.; Kim, H.; Kamat, V.R.; Kang, L. BIM-Based construction scheduling method using optimization theory for reducing activity overlaps. *J. Comput. Civ. Eng.* **2015**, *29*, 04014048. [CrossRef]
107. Faghihi, V.; Reinschmidt, K.F.; Kang, J.H. Construction scheduling using genetic algorithm based on building information model. *Expert Syst. Appl.* **2014**, *41*, 7565–7578. [CrossRef]
108. Li, N.; Becerik-Gerber, B.; Krishnamachari, B.; Soibelman, L. A BIM centered indoor localization algorithm to support building fire emergency response operations. *Autom. Constr.* **2014**, *42*, 78–89. [CrossRef]
109. Nour, M.; Hosny, O.; Elhakeem, A. A BIM based approach for configuring buildings' outer envelope energy saving elements. *J. Inf. Technol. Constr.* **2015**, *20*, 173–192.
110. Liu, H.; Al-Hussein, M.; Lu, M. BIM-based integrated approach for detailed construction scheduling under resource constraints. *Autom. Constr.* **2015**, *53*, 29–43. [CrossRef]
111. Won, J.; Lee, G. How to tell if a BIM project is successful: A goal-driven approach. *Autom. Constr.* **2016**, *69*, 34–43. [CrossRef]
112. Kim, H.-S.; Kim, S.-K.; Kang, L.-S. BIM performance assessment system using a K-means clustering algorithm. *J. Asian Arch. Build. Eng.* **2021**, *20*, 78–87. [CrossRef]
113. Succar, B.; Kassem, M. Macro-BIM adoption: Conceptual structures. *Autom. Constr.* **2015**, *57*, 64–79. [CrossRef]
114. Becerik-Gerber, B.; Rice, S. The perceived value of building information modeling in the U.S. building industry. *Electron. J. Inf. Technol. Constr.* **2010**, *15*, 185–201.
115. Suermann, P.C.; Issa, R.R. Evaluating industry perceptions of building information modelling (BIM) impact on construction. *J. Inf. Technol. Constr.* **2009**, *14*, 574–594.
116. Li, H.; Xiong, L.; Liu, Y.; Li, H. An effective genetic algorithm for the resource levelling problem with generalised precedence relations. *Int. J. Prod. Res.* **2018**, *56*, 2054–2075. [CrossRef]
117. Rausch, C.; Sanchez, B.; Esfahani, M.E.; Haas, C. Computational algorithms for digital twin support in construction. In Proceedings of the Construction Research Congress 2020, Tempe, AZ, USA, 8–10 March 2020; pp. 191–200. [CrossRef]

118. Lilis, G.N.; Giannakis, G.I.; Rovas, D.V. Automatic generation of second-level space boundary topology from IFC geometry inputs. *Autom. Constr.* **2017**, *76*, 108–124. [CrossRef]
119. Seghier, T.E.; Ahmad, M.H.; Lim, Y.-W. Automation of concrete usage index (CUI) assessment using computational BIM. *Int. J. Built Environ. Sustain.* **2019**, *6*, 23–30. [CrossRef]
120. Mukkavaara, J.; Olofsson, T.; Jansson, G.; Sandberg, M.; Elgh, F. Structures for supporting BIM-based automation in the design process. In Proceedings of the 35th International Symposium on Automation and Robotics in Construction (ISARC 2018), Berlin, Germany, 20–25 July 2018; Volume 35, pp. 362–368.
121. Leśniak, A.; Górka, M.; Skrzypczak, I. Barriers to BIM implementation in architecture, construction, and engineering projects—The polish study. *Energies* **2021**, *14*, 2090. [CrossRef]
122. Lim, Y.W.; Seghier, T.E.; Harun, M.F.; Ahmad, M.H.; Samah, A.A.; Majid, H.A. Computational bim for green retrofit-ting of the existing building envelope. *WIT Trans. Built Environ.* **2019**, *192*, 33–44. [CrossRef]
123. Oti, A.H.; Tizani, W.; Abanda, F.H.; Jaly-Zada, A.; Tah, J.H.M. Structural sustainability appraisal in BIM. *Autom. Constr.* **2016**, *69*, 44–58. [CrossRef]
124. Cheng, C.-W.; Cheng, G. Enhancing multi-model inference with natural selection. *arXiv* **2019**, arXiv:1906.02389.
125. Li, J.; Li, N.; Afsari, K.; Peng, J.; Wu, Z.; Cui, H. Integration of building information modeling and web service application programming interface for assessing building surroundings in early design stages. *Build. Environ.* **2019**, *153*, 91–100. [CrossRef]
126. Altohami, A.B.A.; Haron, N.A.; Ales@alias, A.H.; Law, T.H. Investigating approaches of integrating bim, iot, and facility management for renovating existing buildings: A review. *Sustainability* **2021**, *13*, 3930. [CrossRef]
127. Siegele, D.; Leonardi, E.; Ochs, F. A new MATLAB simulink toolbox for dynamic building simulation with B.I.M. and hardware in the Loop compatibility. In Proceedings of the 16th IBPSA Conference, Rome, Italy, 2–4 September 2019; Volume 4, pp. 2651–2658. [CrossRef]

MDPI

Review

Investigating Approaches of Integrating BIM, IoT, and Facility Management for Renovating Existing Buildings: A Review

Abubaker Basheer Abdalwhab Altohami *, Nuzul Azam Haron, Aidi Hizami Ales@Alias and Teik Hua Law

Department of Civil Engineering, Faculty of Engendering, University Putra Malaysia, Serdang, Selangor 43400, Malaysia; nuzul@upm.edu.my (N.A.H.); aidihizami@upm.edu.my (A.H.A.); lawteik@upm.edu.my (T.H.L.)
* Correspondence: gs50517@student.upm.edu.my

Abstract: The importance of building information is highly attached to the ability of conventional storing to provide professional analysis. The Internet of Things (IoT) and smart devices offer a vast amount of live data stored in heterogeneous repositories, and hence the need for smart methodologies to facilitate IoT–BIM integration is very crucial. The first step to better integrating IoT and Building Information Modeling (BIM) can be performed by implementing the Service-Oriented-Architecture (SOA) to combining software and other services by replacing the sematic information that was failed to display elements of indoor conditions. The other development is to create link that able to update static models towards real-time models using SOA approach. The existing approach relies on one-way interaction; however, developing two-way communication to mimic human cognitive has become very crucial. The high-tech approach requires highly involving Cloud computations to better connect IoT devices throughout Internet infrastructure. This approach is based on the integration of Building Information Modeling (BIM) with real-time data from IoT devices aiming at improving construction and operational efficiencies and to provide high-fidelity BIM models for numerous applications. The paper discusses challenges, limitations, and barriers that face BIM–IoT integration and simultaneously solves interoperability issues and Cloud computing. The paper provides a comprehensive review that explores and identifies common emerging areas of application and common design patterns of the traditional BIM-IoT integration followed by devising better methodologies to integrate IoT in BIM.

Keywords: BIM; IoT; integration BIM–IoT; creating database; live data

Citation: Altohami, A.B.A.; Haron, N.A.; Ales@Alias, A.H.; Law, T.H. Investigating Approaches of Integrating BIM, IoT, and Facility Management for Renovating Existing Buildings: A Review. *Sustainability* **2021**, *13*, 3930. https://doi.org/10.3390/su13073930

Academic Editor: Antonio Garcia-Martinez

Received: 6 February 2021
Accepted: 2 March 2021
Published: 2 April 2021

1. Introduction

In the last three decades, a new field in construction and design has been evolving with a revolutionary approach called Building Information Modelling (BIM). BIM is defined by [1] as "a model-based process of generating and managing coordinated and consistent building data that facilitates the accomplishment of established sustainability goals." This definition means that BIM has reached a level to facilitate high-level analysis and evaluations for building by employing techniques such as acoustic analysis, carbon emission, construction and demolition waste management, operational energy use, and water use. In addition, BIM of multidisciplinary data for various analyses could be expressed in a 3D model [2].

BIM takes full consideration of the environmental issues in terms of increasing demand for new and renovating buildings especially in the field of the high-quality indoor environment that has impacted the designing and construction of building. Further, BIM could cover the old buildings by keeping them in a good shape and environmentally suitable for living [3].

The existing building practitioners are facing a series of challenges such as the unavailability of BIM records—the matter that causes very poor assessment and lesser accuracy [3].

Therefore, existing buildings, using BIM technology, can be very productive when renovators are using 3D modeling, information management, and other information gathered from owners or tenants [4]. The other factor of the BIM success is that BIM is a data-rich, intelligent, and object-oriented parametric building modeling tool [5]. It can feed information of various categories into a 3D model. Additionally, BIM innovative development could provide opportunities to support green buildings via employing high-tech programs or devices such as the Internet of Things (IoT) and smart devices. These devices produce high quality live data about the building. The gathered information from different sources can be stored in heterogeneous repositories [6]. Based on the definition of IoT which emphasizes the nature of interconnection of sensing and actuating devices, the ability to share information through a unified framework could develop a common operating picture for enabling innovative applications [7]. The critical importance of IoT lies in allowing sensitive technologies such as sensing, identification, and recognition run by advanced hardware, software, and cloud platforms to draw better sense for how to deal correctly with the renovation and/or demolishing [8]. IoT is an architecture that uses intelligent devices, smart mobile devices, single board computers, different types of sensors and actuators [9].

The facility management industry may be a source of information; however, this information is invaluable in the sense of providing and delivering a timely and professional analysis in addition to the level of consultation support for more effective management services. The integration of information may provide value despite existing barriers and considerations upon combining the managerial building information with the live data [6]. At this point, considering these barriers may stand for a potential approach that supports integrating information and live data academically and industrially [6]. For these reasons, a comprehensive review of emerging areas of application and common design patterns to tackle BIM–IoT device integration in addition to examining the current limitations and predictions [10]. BIM–IoT integration is still in the early stages, which needs serious efforts to achieve a better understanding of the current situation [10]. The benefit for the real-time integration of the environmental and localization data could help in operational construction and facility management by applying the cloud-based BIM platform in the construction and facility management and operation vis two case studies [11].

The combination of IoT and the Lean and Injury-Free (LIFE) construction management may conceptualize the implementation of the topics in existing systems to designing and creating a valid prototypical application in field-typical work settings [11]. Keep in mind that BIM has been relying on data includes building characterization information which is very important in offering information on previous stages of the building's life cycle. The life cycle can be shown through plans either in paper version or digital drawing (Auto CAD), reports, tables, and others. As reported recently by [6], BIM technology could be available in a better platform than thought to be.

The basic rule of BIM in existing buildings relies on delivering or providing information that retrieved from any source that describes the conditions of these buildings or original data. In either case, informative data could be used towards evaluating the renovation process. However, the benefit of this information could be more valuable upon integrating this information in BIM. There are two reasons for this approach: the first reason is that this information came from different sources with a different format, and the second reason, BIM upon integrating data, can organize and prepare them under certain usable format. As an example, a laser scanning cloud provides good information, but it lacks semantic information or geometrical context which makes BIM searching for external sources such as building specification and construction materials. At the time of the laser scanning, data are useful for high-resolution images for spectral and spatial information [12]. The bottom line is the interoperable software programs and database contents of BIM that are featuring the better usage or application of the available information [13].

The above approach can be seen in Figure 1 which shows two parallel applications of BIM: one for new construction while the second is for existing buildings. For new buildings, the BIM process over the life cycle (LC) consists of inception, brief, design to production (case I) followed by maintenance and deconstruction (case II). In case that Architecture, Engineering, and Construction (AEC)/facility management (FM) stakeholders do not employ all BIM processes, an isolated BIM is then created for certain designated single purposes. BIM processes in existing buildings depend on the availability of pre-existing BIM (case II) [14] or created a new IM process procedure (case III). Buildings information for existing buildings is not available in BIM formats, and hence a case study has been proposed to produce 3D CAD models of existing structures using a semi-automated technique [15,16]. Hence, the only way to implement BIM is to utilize manual reverse engineering processes which is costly and time-consuming (case III) [17].

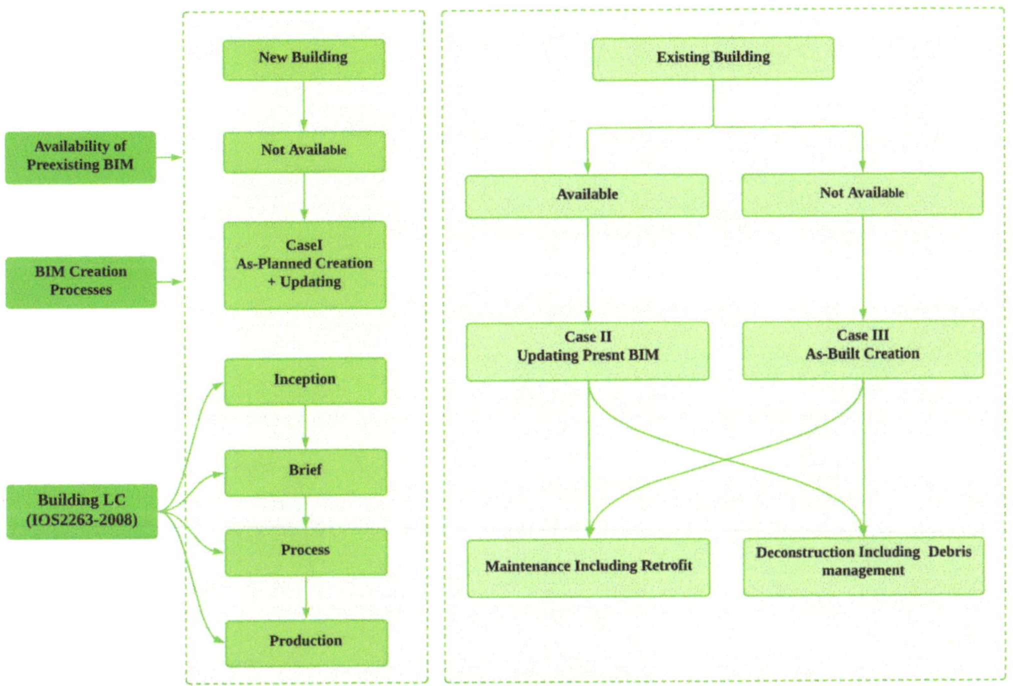

Figure 1. Building Information Modeling (BIM) model creation processes in new or existing buildings depending on available, pre-existing BIM and life cycle (LC) stages with their related requirements [18].

The important step is to utilize information obtained from the high-tech apparatus in the databases of ASCE, IEEE, ACM, and Elsevier Science Direct Digital Library. Meanwhile, the analytical data applied in the construction industry could be saved in the system of Big Data technologies at a very early stage [19]. A collection and relevant analysis of 614 bibliographic records from the science Web database have shown that BIM is mainly developed and applied in the USA, South Korea, and China [20]. Table 1 shows an in-depth analysis of the reviewed papers on the four domains where BIM and IoT were operated. These four domains are operation and monitoring, logistic and management, facility management, and health and safety.

The paper is structured to contain nine sections include the Introduction. In Section 2, this paper discusses the relationship between BIM and the existing building. A discussion for integrating BIM and IoT are discussed in Section 3. Section 4 highlights implementing the big-data principles in integrating BIM and IoT. Section 5 illustrates the approaches to integrating BIM–IoT devices. Section 6 explains the query language. Section 7 contains a summary of previous empirical research. Section 8 highlights the contribution of this study, while Section 9 displays the conclusion.

Table 1. Recent studies on BIM–Internet of Things (IoT) implementation in existing buildings [10].

Construction Type	Reference
Operation and Monitoring	[21–23]
Logistic and Management	[24–27]
Facility Management	[28–31]
Health and Safety	[23–32]

2. BIM and Existing Buildings

It is crucial to outline the flow of the topics in this paper as described in the flow diagram shown in Figure 2. One of the purposes of this flow diagram is to assist the authors in viewing the proposed system in a structured fashion.

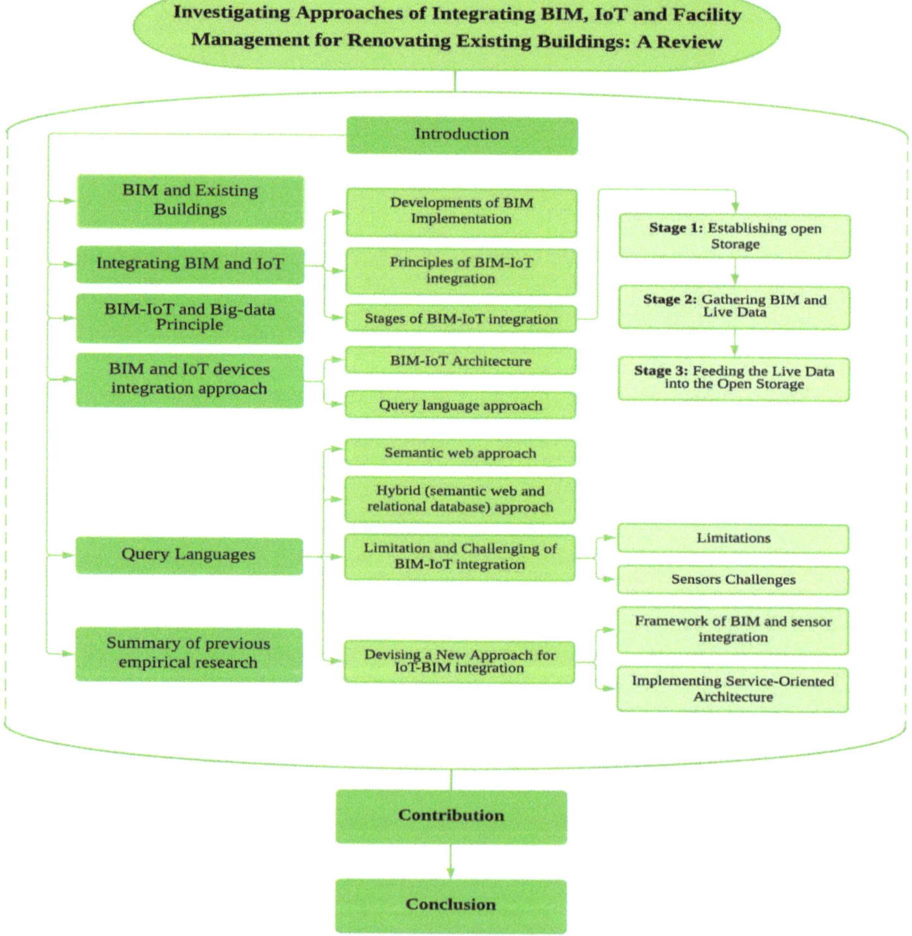

Figure 2. The methodology diagram flowchart.

As stated by [33] that BIM is a new digital and management paradigm that shows great potential for renovating buildings despite critical issues such as missing documentation. Within this concept, the role of BIM is to facilitate the connection with stakeholders including the facility management to collect discontinued or unavailable information towards establishing a database. For this reason, BIM has become increasingly attractive to the management that deals normally with comprehensive and incremental knowledge for accurate assessment of recovering the residual building performances throughout refurbishment and retrofit [34]. This approach has become possible by extending the effective involvement of all technicians who are equipped with multidisciplinary skills besides sharing successful information [35]. The most important step is to integrate BIM and automation systems, which could positively support the quality control throughout diagnosis, design, work execution, and labor savings. Hence, BIM is characterized as a guide tool that controls the flow of work and information and stands on integrating digital archive of geometric, semantic, and topological data, which can be performed in various formats and contents within parametric objects [36]. The customized tools of BIM can manage and examine the variable multiplicity due to query operations and specifically programmed automation algorithms [37].

The involvement of BIM in heritage buildings, as a part of existing buildings, was initially identified as Historic Building Information Modelling (HBIM) [38]. Since then, HBIM has been adopted by many researchers [39]. The previous definition of BIM excludes information management. However, including this information in later BIM definition made an evolution in terms of approaching the Built Heritage Information Modelling and Management (BHIMM) [40].

The existence of state-of-the-art in BHIMM for building refurbishment whether in general terms [41] or for existing buildings [42] has made a very important step towards integrating BIM with other tools. Despite that, more guidelines are still needed to achieve a complete "as-built" model featured with morphology (regular or irregular) of historical buildings hoping to finalize an accepted design of refurbishment and conservation interventions as noted earlier by [43] and recently by [44]. Meanwhile, researchers face a critical activity when BIM and a variety of information through independent and structured methods. Such information can be gathered from various sources such as historically archived documentation and analytical investigations in addition to taking surveys, diagnostics, monitoring, and continuous updating information during designing, execution, or performance assessment [39]. HBIMM can be used successfully in improving BIM ability to include tangible and intangible information for the existing buildings, in general, as reported earlier by [45] and later by [39]. This information could be available to managerial and maintenance stakeholders throughout the life cycle. Further, fortifying this information with real-time updates may help to centralize models by installing integrated monitoring systems [46,47]. The state-of-the-art was drawn up to achieve the following four purposes: a critical analysis of potentialities of HBIMM, formalizing the first attempts of HBIMM application, proposing a consolidated methodological flow, and suggesting diagnosis tools for future developments within an automation-based framework.

3. Integrating BIM and IoT

Reference [48] defined data integration as "the combination of data from different sources with unified access to the data for its users". To achieve the goal of data integration, many researchers have proposed methods and models. However, integrating data in this study is limited to integrate the building information with the live data to achieve better facilitation to the maintenance of the buildings. It is important to note that the result of data integration so far is showing inadequate processes because of challenges faced by BIM and other building information management technologies as reported recently by [49,50]. In this section, a description of the foundations of data integration is presented, followed by explaining the challenges and barriers, and finally, the three stages necessary to achieve BIM and IoT integration.

3.1. Developments of BIM Implementation

Generally, modification, altering, developing a new concept must come with a reason. Integrating IoT in BIM can be seen through the progressive developments of BIM as influential technology despite the appearance of some shortcomings or barriers [51]. BIM historically has become the core of information management in the Architecture, Engineering, and Construction (AEC) industry, operation, and maintenance (O&M), and facility management (FM) [52]. Another benefit of emerging BIM is that it allows stakeholders to exchange and manage information about building throughout the building lifecycle as early reported by [53] and then by [54]. In addition, it has been reported that BIM could become an influential tool for analyzing energy usage, defect detection, firefighting, renovation and demolition, and safety in the facility [55]. Among these areas, the adoption of BIM in the AEC industry was very well established, whereas BIM has received growing attention for O&M and FM [30]. The implementation of BIM in Architecture, Engineering, Construction, and Operation (AECO) was limited to include only new projects [56].

It is a fact that BIM was created to be applied heavily in new construction projects; however, there is another parallel movement to utilize BIM in existing buildings where BIM has not existed [57]. The sematic approach could rich the creation of BIM to face challenging, complex, and expensive considering [57]. In addition, generating BIM has become complex since it relies on the level of detail (LoD), intended use, interoperability, and functional issues [58]. Accordingly, implementing BIM for the existing building has yet to be fully realized.

In construction, it is a well-known fact that the O&M period is the longest in the building lifecycle. In this period, all buildings require a monetary expense for labor, materials, maintenance, and renovation that makes this period the costliest [59]. It was reported by [60] that 85% of the total project cost is spent on O&M activities. Meanwhile, O&M information is not accurate and can be described as fragmented due to manual and tedious reporting method. The following barriers hinder attaining an effective O&M:

- Under scientific and technological developments, building systems have become increasingly complex due to utilizing sophisticated technologies, security issues, and sustainability [61,62].
- Maintenance and repair of the life cycle of O&M and FM result in injuries caused by falling, electric shocks, crushing, and other workers and facility users.
- Retrofitting existing buildings poses a great technical challenge such as energy interaction of the audit system, building performance and risk assessment, and energy savings [63,64].
- According to [65] and recently to [66], a large portion of total energy worldwide has been spent on the current operations of the existing building. Hence, utilizing energy efficiently is indeed a prime concern in reducing stress on the energy system and benefits the environment. In this sense, BIM can significantly improve energy analysis through simulation [67].
- The fragmented O&M causes energy simulation very difficult leading to a possible error-prone decision in the renovation along with issues such as safety, repair, and maintenance [68].

According to the above analysis, the more and accurate O&M information the best outcome for the building's lifecycle. Hence, BIM, digital environment, and 3D visualization capabilities, enabling storage, sharing, and integration of information are essential for all stages in the renovation of existing buildings [69]. The successful implantation of BIM during design and construction using the digital database is very necessary for management and stakeholders. However, it is very important to note that using the same BIM in O&M and FM could cause limited inconsistencies and the possible absence of the uncounted number of information necessary for O&M and FM [70].

Moreover, like advanced technology, the Construction Operations Building Information Exchange (COBie) was created to eliminate the interoperability issues between the model and construction operated by FM [46]. Briefly, BIM integrated into O&M and FM for the existing building still needs updating.

3.2. Principles of BIM–IoT Integration

Technology has been found very useful in enriching BIM with information that helps in adopting renovating existing buildings, such as 2D floorplans and elevation drawings. 2D can semantically enrich BIMs with proper and useful geometry for energy simulation for daily operations in addition to flexible building information from the facility management to formulate an informative model approach [71]. The input/output (IO) data has been utilized to quantify the embodied energy intensity of residential and commercial sectors in the USA [72]. In another approach, the optimization of the multi-objective algorithm could be implemented to improve the building stock energy efficiency, sustainability, and comfort, while efficiently allocating the available budget to the buildings [73].

This process requires expert knowledge to handle the core steps to deal with challenges that occurred in handling occlusions/uncertain data. In this case, the conversion process has become cumbersome as BIM data increased to a level of detail (LoD) which, then, requires using different techniques to capture imaging, 3D scanner, Ground Penetration Radar (GPR), and 2D scanned plans [52]. The missing BIM of most existing buildings adds restrictions for constructing BIM using high-tech devices.

On the other side, employing a BIM-based Data Mining (DM) approach for detecting improper records may lead to construct a BIM database. This is an introductory step towards transferring the database into a data warehouse where the DM method shows useful information from the BIM [74]. The BIM-oriented approach produces 3D models based on gathering data for geometrical and non-geometrical information related to various several themes.

These themes include historical documents to monitor other data that was created by reflecting the shared parameters for the ontology domain. The result of this approach is structuring data in a machine-readable format by converting needed data and set this data in a domain [75].

3.3. Stages of BIM–IoT Integration

The integration of BIM and IoT devices has been widely considered as a powerful paradigm processing aiming at improving construction and operational efficiencies. This integration is meant to serve the rapid expansion of IoT sensor networks and to cope up with these developments to achieve establishing high-fidelity BIM models that, ultimately, provide numerous applications. There are two important issues about BIM–IoT integration. The first issue is that this integration is still in its early stage. The second issue is that there is more than one approach for this integration depending on the reason and the limitation of such integration [10].

In one of the several BIM–IoT integration methods, utilizing the existing BIM tools' Application Programming Interface (APIs) and relational database was adopted as shown in Figure 3. This integration requires storing the time-series data from sensor time-series data stored and update it in a relational database in forms such as SQL server database or Microsoft Access. In addition, exporting BIM models constructed in BIM tools such as Revit into a relational database. The next step is to define a database schema to clarify the relationship between virtual objects and physical sensors. It is important to highlight that such integration requires a two-way importing and exporting of a relational database and BIM model. Lastly, BIM–IoT integration should process queries of sensor data through custom-built API in the form of the graphic user interface (GUI) and direct query over SQL database [10].

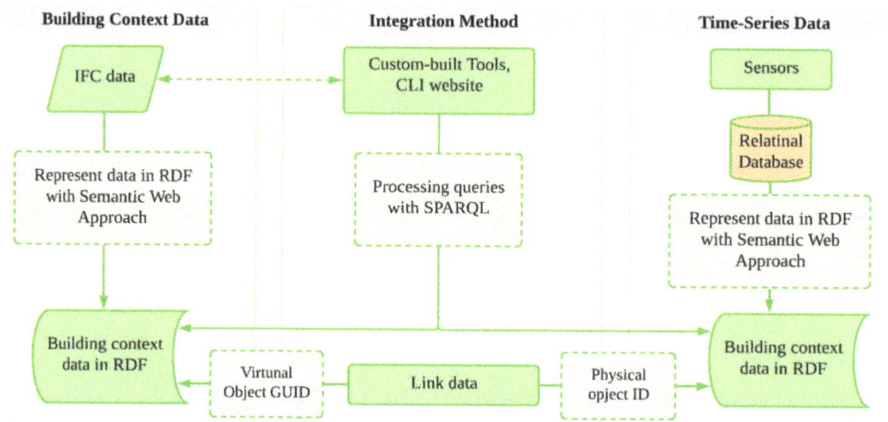

Figure 3. Integration of BIM tools and the relational database [10].

As mentioned above, BIM and IoT integration can be performed differently using a systematic method throughout three stages. The foundation of this systematic integration was based on the challenges that are facing BIM–IoT integration. It seems that tackling these challenges cannot be performed directly; seemingly, it needs a detailed process with definite stages. This approach suggests collection, analyzing, building information with the live data captured from various IoT and smart devices and sensors. The proposed process includes three stages explained as follows:

3.3.1. Stage 1: Establishing Open Storage

The first stage of the proposed process model is to facilitate storage in which the integrated, qualified building information, and live data are stored according to the construction industry standards such as ISO 16739, ISO 12006, ISO 29481, and, in some cases, the European standards for the construction industry. These standards require structural and semantic requirements with specific data systems. In this regard, the data storage should bed initially structured and well defined [76]. The final structure of the open storage could be completed after having the following two stages completed and assessed.

3.3.2. Stage 2: Gathering BIM and Live Data

Stage Two includes proposing various sub-processes aiming at capturing the building information and the live data from various sources such as reports and IoT devices and sensors. Firstly, digitalizing the building information taken from sources such as architectural, mechanical, electrical plans as well as the project reports. It is expected that building plans contain important and basic information about the buildings in terms of spaces and specific devices. In addition, there is another source of information that comes from mechanical development, electrical design and consumption, and structural systems. At this level, the basic information required is gathered and ready for the digitalizing process. As a connection to the coming sub-stage process, the digitalized data and information should be linked to the plans and the project reports [76].

The Stage Two-second sub-process includes detecting all installed IoT devices and sensors in the building by updating the information of these devices. The detection is normally carried out by laser scans followed by capturing images and video records. The outcome then develops a list of devices in cloud data forms and converts them to objects. Stage two could benefit from the data stored according to the previous Stage one. At the end of Stage two, it is important to note iterative processes of feeding the live data into the storage is explained in the following Stage Three.

3.3.3. Stage 3: Feeding the Live Data into the Open Storage

Stage Three is the last stage where the proposed process model will be transferred into qualified live data for storage. Collectively, captured data from the IoT and smart devices must be stored according to the software of the databases. This critical step requires a well-defined interface that enables transferring data to the storage. The issue of live data mentioned earlier in Stage 2, the installation is carried out concurrently with the installed devices in the building spaces which can be transferred to the corresponding fields in the storage. In all stages, the quality of the stored data should pass the control test before getting ready for use [76].

4. BIM–IoT and Big-Data Principle

A big database is one of the perquisites of the digital revolution [77]. There are several sources of the big database with diverse disciplines including initiation, modeling, engineering drawings, and the facility life cycle. In addition, BIM has been proposed to involve the technology that captures multi-dimensional CAD information supporting multidisciplinary stakeholders [23,78]. The BIM data encodes 3D geometrical, computed intensively with graphic and Boolean computing, compressed in various formats, and intertwined [79]. Seemingly, the diverse data within BIM models gradually continued beyond the end of the facility. For example, the size of the BIM files with designing a three-story building could hit 50 GB [80]. This data can be in any form or shape and constitutes an intrinsic value to industrial performance. The evolution of the advent of embedded devices and sensors was a good approach to facilitate and generate massive data during the operational and maintenance stage leading to the creation of a Big BIM Data system [81]. Consequently, the vast accumulation of the BIM data has pushed the construction industry to enter the Big Data era. Big Data is characterized by three attributes that are commonly known as 3V's. These three V's are the volume (terabytes or petabytes); the variety (heterogeneous formats like text, sensors, audio, video, graphs, and more); and the velocity (continuous streams of the data) [82]. The formats of the construction data include DWG (drawing), DXF (drawing exchange format), DGN (Design), RVT (Revit), ifcXML (Industry Foundation Classes XML), ifcOWL (Industry Foundation Classes OWL), DOC/XLS/PPT (Microsoft format), RM/MPG (video format), and JPEG (image format). Additionally, the nature of construction data is natural dynamics due to the streaming properties of data sources such as sensors, Radio Frequency Identifications (RFIDs), and BMS (Building Management System).

At this level of understanding, it is important to remove ambiguity between Big Data Engineering (BDE) and Big Data Analytics (BDA). Firstly, the BDE domain presents the need for analysis of the data storage and processing activities. In addition, Big Data Analytics (BDA) relates to the tasks responsible for extracting the knowledge to drive decision-making [83,84]. According to [85], the nature of BDA is to discover the latent patterns that is buried inside Big Data. As such, it becomes plausible to transform the future of many industries through data-driven decision-making. For future expectations, identifying, understanding, and reacting to the latent trends promptly could show a competitive edge in this hyper-competitive era.

The big data is controlled by Mappers and Reducers (MR) and to be represented by a processing model where the mapping is processed first followed by reduction as shown in Figure 4. The analytical tasks in MR are written as two functions according to [86] and recently to [87]. It can also be noted that there is an intermediate stage sandwiched between MR's steps. In principle, in the mapper stage, data read, processed, and then used to generate intermediate results. The output of the mapper staged is treated by reducers to finalize results that are stored back to the file system using a platform called Hadoop. The typical Hadoop cluster contains several MRs working simultaneously within a powerful model for batch-processing tasks. The challenge to this model is that the applications require real-time, graph, or iterative processing. This challenge has been encountered by a recent version of Hadoop (Hadoop 2.6.0). In addition to Hado's models, there is another model called the Yet Another Resource Negotiator (YARN) which is used to utilize Hadoop towards a new approach for developing a Big Data platform. Under this approach, the service is run by MR over YARN, while YARN handles scheduling and resource management to produce Hadoop that is suitable for implementing innovative applications [19]. YARN is the technology that is designed for cluster management and is one of the key features in the second generation of Hadoop, the Apache Software Foundation's open-source distributed processing framework.

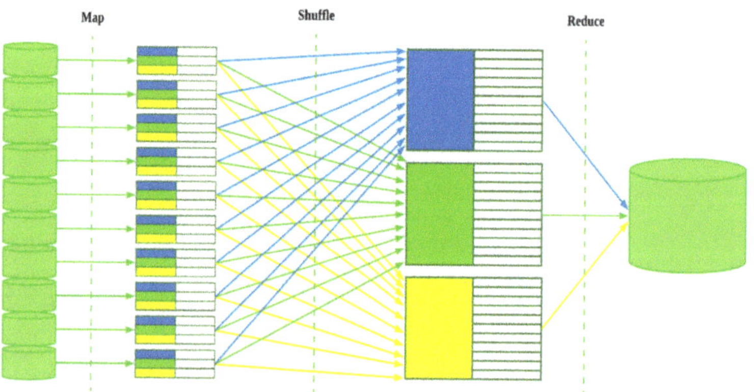

Figure 4. MapReduce processing—an overview [88].

In another development, a new technology called Directed Acrylic Graphs (DAG) has emerged to better handling big data by modifying the model shown in Figure 5 into a map-then-reduce style. The most functioning element in the DAG model is called the Spark, which was originally proposed by [89] and then elaborated by [90]. Spark is characterized by its in-memory computation and high expressiveness [91]. Based on these capabilities, Spark has become a natural choice to support two components of Big Data in iterative and reactive applications [92]. Regarding the speed, Spark has been reported having ten times faster than MR on disk-resident tasks and a hundred times faster for the memory-resident task [93].

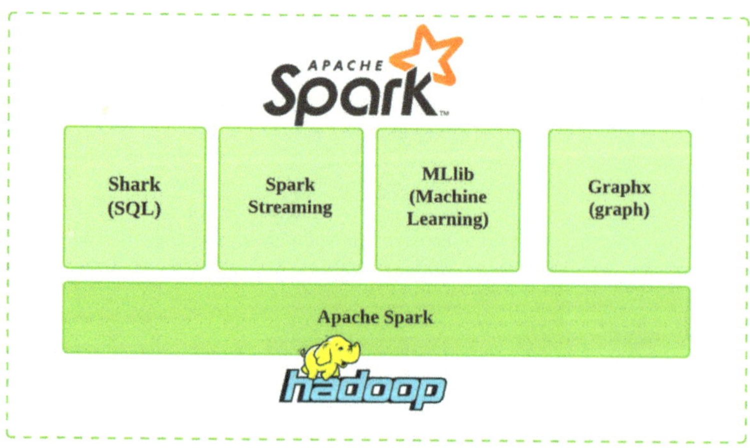

Figure 5. Spark and related technology stack [94].

A schematic representation of the architecture and the steps involved in the analytics process of Hadoop and Spark frameworks is shown in Figure 6. The two frameworks involve a new form of MR called elastic MapReduce (EMR), which transfers MR to a system comparable to the cloud [95]. EMR can also run processing frameworks such as Apache Spark and HBase on Hadoop clusters for batch processing, querying, streaming, and machine learning.

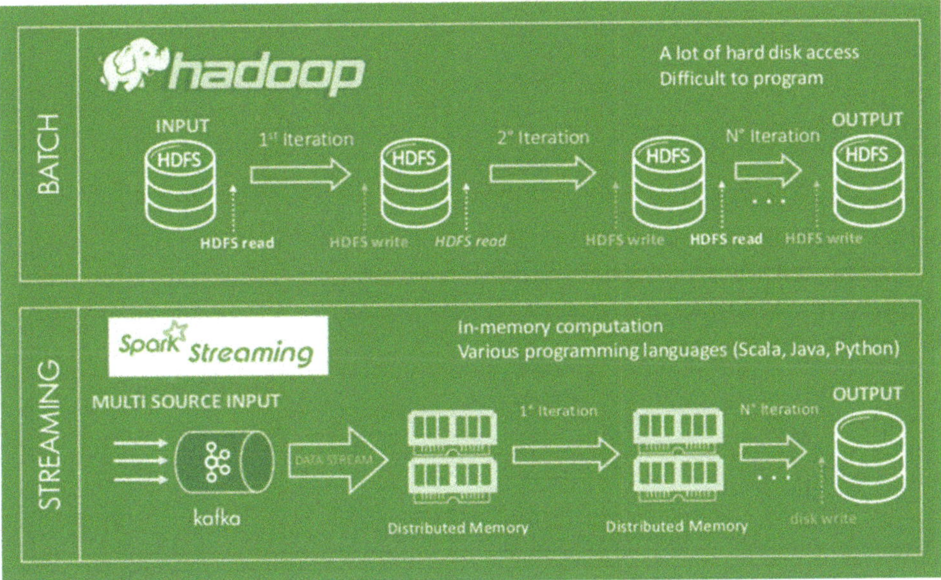

Figure 6. Comparison between Hadoop and Spark [95].

5. BIM and IoT Devices Integration Approach

Integrating two or more systems is a very common practice in technology. In this paper, integrating two very advanced technologies, BIM and IoT, requires attention to the following components:

- BIM acts as a data repository for contextual information including building geometry while IoT acts as static and soft information that is gathered from occupancy patterns. In addition, IoT is responsible for scheduling data like social media, feedback originated from FM and occupant interactions, and external sources of information such as weather forecast and financial pricing [96].
- The continuous sensor readings and traditional time-series data (known as time-series data) are stored in a well-structured relational database queried effectively under Structured Query Language (SQL) [97,98]. The time-series data is provided by sensors while API is responsible to move data belongs to BMI models constructing tools. The database is to clarify the connection of a two-way connection between virtual objects and physical sensors where data is passing between the database and BIM using APIs [99].
- The third component is about integration between contextual information and time-series data.
- Integrating BIM and IoT requires an interface technically known as Application Program Interface (API) to correlate data from the sensor and BIM model with evolving database and the two-way importing/exporting data, and queries processing [100].

5.1. BIM–IoT Architecture

Figure 7 shows a proposed architecture where the data is collected from various heterogeneous IoT sources. The data then undergoes anonymization followed by processing and analysis using the artificial intelligence (AI) technique. The outcome of this data analysis is mainly towards optimization of some parameters of interest aiming at achieving a desired security and performance objectives. The process is carried out in terms of available resources throughout six phases: phase 0 through phase 5. Phase 0 is to perform security parameters configuration by employing cryptographic algorithms to cater to the confidentiality, integrity, and authentication requirements of the applications and systems using this specific device. Phase 1 is about data sensing and reporting where main IoT classes are distinguished while they will be utilized wither by being wearable devices or IoT appliances. Phase 2 deals with data aggregation and relaying where both need a decision which is target-dependence known as IoT gateway. The third phase (III) involves cloud-based data analysis for the data sent by IoT-gateway. Data analysis is carried out with preserving privacy since this information is made for the general population through Google.

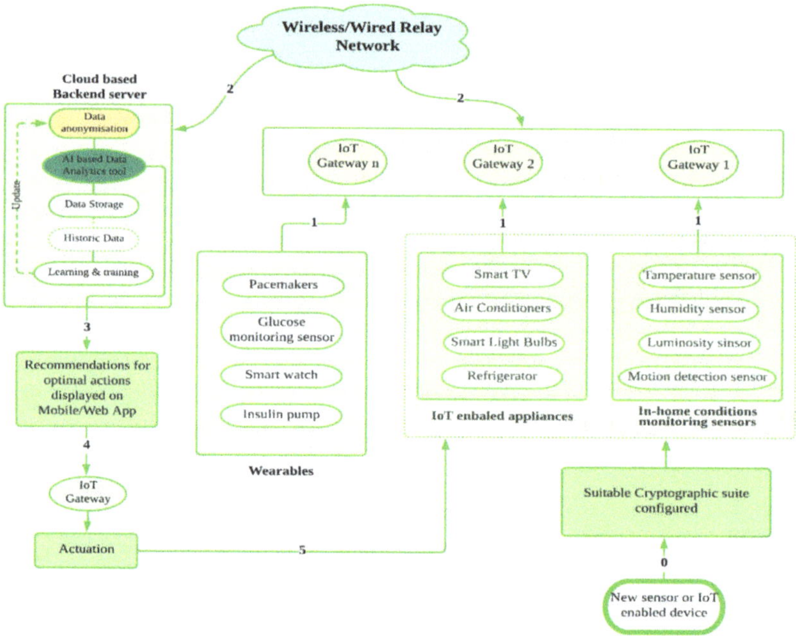

Figure 7. Architecture of a secure and privacy-preserving IoT-based sensing and actuation system [101].

At this level, phase IV is implemented to optimize decision delivery which represents the measure for actuation the system via IoT gateway in terms of mobile or web applications. The last phase is called the actuation phase where instructions should be made to update the configuration file. The degree of actuation could range from changing the upper and lower limits of a given parameter so that reading beyond these limits to trigger an alert and setting conditions under which certain tasks are performed [101]. The phases mentioned above facilitate transferring data into a query-able database. Moreover, the stored data of the traditional building management system can be moved to a well-structured and effective SQL query. This is a fundamental step for binding time-series data with BIM. The linking of

BIM data can be performed using SQL query-able where connecting virtual sensor objects with physical sensors can be done via a Globally Unique Identifier (GUID) [102].

5.2. Query Language Approach

There are several query languages approaches to execute a natural language query (NLQ) and to form a query that is understood by the machines and saved in a database that can be displayed using the graphical user interface (GUI) as shown in Figure 8 [103]. This approach is meant to create a new query language based on processing time-series data query sensor data instead of using SQL as reported by [104]. The authors proposed a domain-specific query language called BIMQL used to modify selected queries from industry foundation classes (IFC)-based BIM models. BIMQL allows object selection and attributes based on arbitrary properties as in IfcSensor despite the limitation in query real-time sensor data [10].

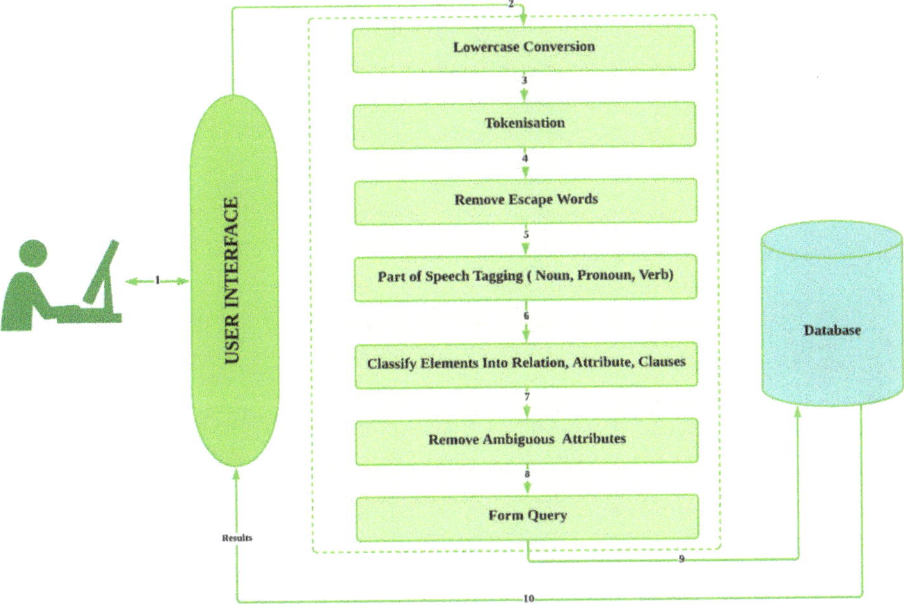

Figure 8. Process control flow diagram [103].

6. Query Languages

6.1. Semantic Web Approach

In 2000, [105] proposed the CommonKADS methodology which was then introduced to web-oriented knowledge engineering and management. CommonKADS helps in constructing, representing, and accessing appropriately the Web/Grid which becomes recognizable, sharable, and reusable by humans and machines. The methodology of several activities for managing Grid knowledge is shown in Figure 9.

The system is organized in seven phases, i.e., Application Analysis, KM Analysis, Ontology Development, Semantic Annotation, Service Development, Testing and Evaluation, and System Integration. For each of these seven phases, there is a list of tasks called activities which, then, were described by outcomes [106]. The outcome includes a wide variety of activities that range from being attributed to the documents, users, services, testing, and ontologies. In BIM, the task of the sematic web approach is confined to gathering data throughout across building's lifecycle in modern AEC processes is obtained from different sources of data such as building geometry, topology, IoT sensor, and BIM geospatial information. Gathering and stores data does not help in offering a full picture. Hence, integrating BIM and semantic web technologies results in acquiring heterogeneous data sets which should be presented in Resource Description Format (RDF) for using and sharing. Moreover, BIM ontologies (IfcOWL) and Smart Appliances Reference ontology (SAREF) can be performed using Semantic Sensor Network (SSN) to successfully approach BIM and IoT device integration [10].

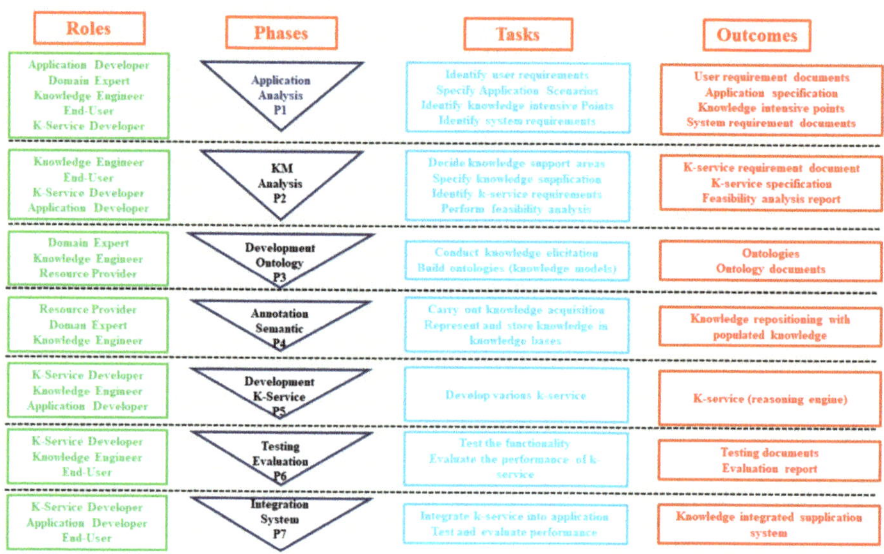

Figure 9. The methodology for the Semantic Web based approach [106].

Based on the discussion of semantic Web-based, an ontology framework named as Onto FM has been proposed to show how monitoring intelligent sensor-based building could fit in this approach [107]. As a complementary process, IFC was introduced to show converting the building geometry into Web Ontology Language (OWL) while using SPARQL to conduct ontology queries. Another study showing a more explicit example of using RDF-Cloud-Linked Data to integrate cross-domain building data from [108]. In addition, other query languages of SQL and XQuery could help in translating SPARQL queries as proposed by [109]. The importance of the monitoring system for real-time data can be seen in other applications fields such as healthcare system [110] and, more broadly in Big Data studies that involve machine learning models, data preprocessing, missing data imputation, and site reliability engineering (SRE) [111].

The advantage of this approach could be seen in linking a homogeneous data format with the cross-domain data. This approach faces some challenges such as storing most of the time-series sensor data in a well-structured and relatively mature relational database, possible data duplication. As a result, converting sensor data in the RDF format leads to RDF inefficiency [109]. The benefit of this approach is to widen the transformation of the knowledge of the semantic web. This benefit represents the possibility of achieving the real concept of interlinking IoT with the Internet via a unified and concise framework provided avoiding complex and heavy data transformation.

6.2. Hybrid (Semantic Web and Relational Database) Approach

Hybrid and integration are two different approaches. As hybridization is "working, organizing, or doing something that is composed of elements of two separate systems", the integration is "the act of bringing together smaller components into a single system that functions as one" [112]. In the hybrid approach shown in Figure 10, the Semantic Web and relational databases are used to store cross-domain data. Hybridization can be performed through feeding contextual information (building context data, sensor information, and other soft building information) into RDF format using a semantic web approach. Then, retaining sensor (time-series data) in a relational database. The mapping contextual information with time-series data is the last step which can be referenced in terms of sensor ID described in RDF [109].

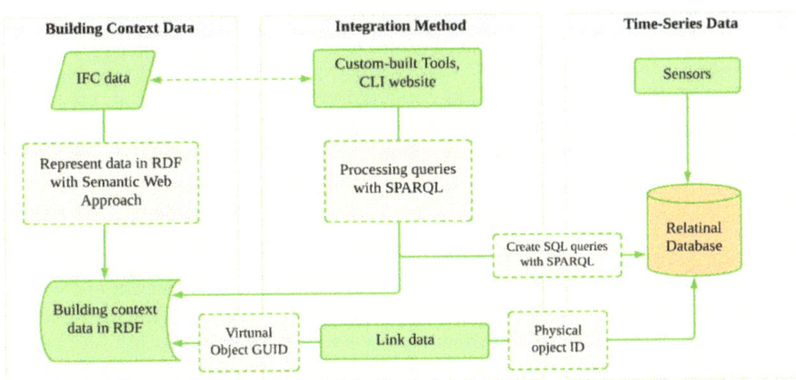

Figure 10. Hybrid approach combining semantic web and relational database [10].

The hybridization method can be performed by converting static sensor information into RDF format to be stored in a relational database and maintained its original form and cross-referenced with SSN ontology to maintain an appropriate platform and format. However, integrating web-hybrid data with BIM still requires modifying sensor data by developing an actuator infrastructure [113]. Hence, any solution is associated with considering integration of the heterogeneous data sources that require building semantic model in RDF format that can be uploaded to a SPARQL server.

The hybrid approach is characterized by involving various data sources retained in the original platforms without influencing the interlinking process. For this reason, hybridization is much effective in storing time-series data while keeping them flexible during building contextual semantic web approach and query. Meanwhile, hybridization has some advantages such as timesaving during storing and duplication, the available size for storing RDF data, better performance, and effective usage of the query language [109]. This approach suggests that it is a very promising method in facilitating IoT integration and it might be suitable for other types of projects that do not require data conversion.

6.3. Limitation and Challenging of BIM–IoT Integration

6.3.1. Limitations

BIM and IoT are new technologies; however, integrating IoT and BIM in construction engineering is another, yet, much-advanced technology. This integration offers the setting of data originated from BIM and IoT as complementary for creating a project. As explained in the preceding sections that BIM and IoT can be integrated; however, this integration has some limitations or restrictions [114].

Individually, the BIM model limitation is related to approaching the component level with high-reliability representations. Generally, BIM models are a very good source for incorporating geometry, spatial location, and a scalable set of metadata properties which, collectively, provide a high-reliability operable dataset. The operable dataset offers suitable conditions for building design by incorporating the spatial organization as a set of virtual assets [115]. IoT data, on the other side, enhances the information set due to employing real-time and recordable status taken physically from real construction operations.

The sensors, sensitive parts of high-tech devices, are a reliable source of sampling potential information from sensors. The data sampling includes three sectors: measurements of the physical building, weather information from official records, and the information from high-tech devices [116]. The characteristics of the data obtained by sensors are series streams distributed over time and frequently at a high-level. The stored IoT and BIM data is accessible through manual interfaces of proprietary systems or programming APIs associated with these applications.

The database provides accessible data export to the systems via open standards that regulate both BIM and IoT fields [10]. The other application of the adoption of BIM and IoT devices was in aspects such as energy management, construction monitoring, health, and safety management, and building management. It is important to note that implementing BIM–IoT is still in the early stages and most studies are in the conceptual and theoretical domain [117]. IoT integration in BIM cannot be considered as merely involving IoT devices (sensors or actuators), but the goal is to interconnect IoT devices to gather the information that can be shared on the Internet. To challenge the system design, a novel robust framework for the energy scheduling of a residential microgrid is connected via smart users [118]. The main current problem of BIM–IoT integration is the difficulty of sharing this information across the Internet under a unified framework [10]. The situation now is that the integration of BIM and IoT device is scattered and need more research to be sufficiently matured in terms of patterns, issues, and opportunities [119]. The smart energy management system has been applied in construction in terms of controlling both heat and electricity using the principle of integrating various types of flexible appliances as well as hybrid energy appliances [120].

The integration of BIM and IoT may provide a source of information that can impact facilitating BIM as a result of installing the smart-nature IoT in the building's spaces. The results could provide very important information concerning the status of both the spaces and devices. The facility management (FM) organizations may benefit from this live data to deliver more added value services to BIM–IoT integration [121]. The possible success of BIM–IoT integration, however, still face barriers to benefit from the live data concept. In this case, two barriers may arise; a major barrier is originated from the fact that the live data are still scattered across heterogeneous storage, while the second barrier comes from a situation that appeared as fragmented systems scattered across space planning, the maintenance helpdesk, the building management system, and the facility management. The implication of these two barriers is complicating the efforts and causing inaccessible data due to the non-standardization data input and the lack of standard processes and procedures for capturing and recording the building information [10]. As such, the absence of standard methods or procedures for capturing data results in forcing FM staff to develop methods and procedures that do not necessarily coincide with the nature of BIM and IoT integration [122].

The absence of combining building the space information and the live data may complicate the task available for FM companies. Hence, research equipped with suitable surveys found in literature aiming at exploring the origins and impacts of the barriers mentioned above on the efficiency of the services. Reference [123] underlined that FM activities require both accuracy and accessibility of the created data in the design and construction stages. However, [124] pointed out the importance of preparing management and organizing competencies for the importance of the FM activities to the managerial procedure.

Meanwhile, the explored barriers can be eradicated hoping to make possible improvements to enhance the overall efficiency of the staff activities and minimizing the time spent on individual tasks. Furthermore, eradicating barriers could lead to enhance the performance of the staff by reducing the required time for establishing the accuracy of the available building information. The saved time may be used to improve the task at hand such as monitoring the use of energy. Then, the overall improvement of lifecycle management can be achieved with a reduction in the operational costs. The benefits of the proposed solution are described in the form of practical use cases [6].

6.3.2. Sensors Challenges

Generating BIM for existing buildings is no easy task; it is very complicated and expensive due to the current and emerging challenges in preparing the required data, modelling, and processing semantic memory [125]. In addition, the BIM's level of detail (LoD) may add more complexities to O&M and FM functionalities [57,88].

The generation of as-built information for creating BIM requires the acquisition technologies which rely on Digital Photogrammetry (DG), Terrestrial Laser Scanner (TLS), and Ground Penetrating Radar (GPR), which, collectively, provide the geometric information of the object. The role of DG is to capture still images and then expressing them in 3D point clouds. Moreover, TLS is to estimate how far the scanner from the target by using multiple points of amplified light. Lastly, GPR equipped with a high-frequency radio signal can infer the location of embedded objects. According to [126], as-built data acquisition requires involving on-site labs furnished with manual or visual measurement assessment. The main challenge of DG, TLS, or GPR shows high-error and requires a long time. Hence, with the poor reliability of these techniques, significant uncertainty is created for decision-makers [127].

To minimize the error created by DG, TLS, and GPR, a series of precautionary steps should be considered. For DG, it is preferable to use simultaneous two 3D cameras to detect the coordinates of the object. The data available creates at least two converging lines to identify a point in space by using several inputs representing coordinates (X, Y, and Z) and the angle of rotation of the cameras (ω, ψ, and κ). Despite this precautionary step, several problems and challenges still exist such as the sensitivity of DG to the changing light conditions during the daytime by introducing shadow which affects the alignment of photos. The other problem is occlusion and the noise in image sensors [128]. The third challenge is the inability of the cameras to provide an absolute scale for distances [129].

Secondly, regarding TLS, the emission of pulses of light to the surface of the object of interest suffers from reflection captured by the sensors causing an error in the derived distance [130]. Hence, replacing old TLS with a more sophisticated TLS can capture more features to better estimate the distances [131]. However, this technique is not suitable for long ranges because of expanding the spot which creates multiple returns causing problems with partial occlusions. Lastly, GPR has been used to identify the location of buried utilities [132] and inspect concrete structures [133].

However, during emitting light to penetrate objects, there will multiple reflections that affect the quality of the images and the measured distance. GPR is limited to detect objects up to 30 m depending on the frequency of the radio signal used. A solution has been proposed to eliminate the multiple reflections was proposed earlier using coupled antennas [132]. Figure 11 shows the employed methodology which includes different non-destructive testing (NDT) techniques for characterization of the outdoors, indoors, and internal structure besides the elements involved.

The process combines all the 3D as-is structure matters such as geometry and materials and the irregularities that affect the structure or the used materials. The external surface geometry and its spectral properties were determined using a camera mounted on a unmanned aerial systems (UAS). The visible pathologies of the façade were also identified from RGB (red, green, and blue) imaging. To complement this information, a GPR system was employed to examine the façade interior (through-the-thickness) while infrared thermography images (IRT) was applied to the nearest sub-surface. All the gathered information, properly georeferenced, was fed to a 3D model obtained by a TLS technique and supported by a BIM model elaborated in Autodesk Revit. Information obtained with other techniques (either NDT or more intrusive) can obviously also be integrated into this BIM model [134].

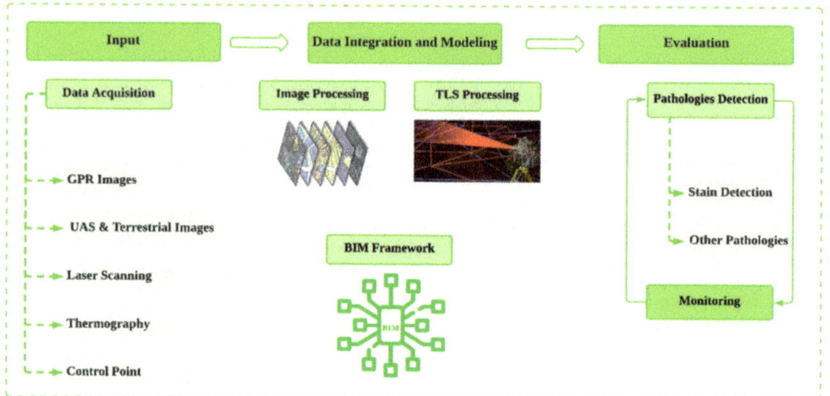

Figure 11. As-built information for creating BIM using Digital Photogrammetry (DG), Terrestrial Laser Scanner (TLS), and Ground Penetrating Radar (GPR) [134].

6.4. Devising a New Approach for IoT–BIM Integration

6.4.1. Framework of BIM and Sensor Integration

An attempt was made to formalize a theoretical framework to integrate information theories with knowledge management. The first item in this formalization is establishing an end-user conceptual framework of BIM and sensors in which the knowledge layer is located from data provided by occupants [135]. The more visionary descriptions from the end-users could identify the best connection of IoT or enhance the ability to collect and analyze data towards sharing information across platforms [7].

In a previously conceptualized framework, analytical activity is not thoroughly explicated [136]. In addition, integrating data from the warehouse, for example, was the only part mentioned without illustrating and emphasizing the analytic functionality of this layer. This could be resulted in by the complication of this new integration. The systematic approach for the conceptual framework is shown in Figure 12. There are four layers that satisfy the pyramid hierarchy of [137].

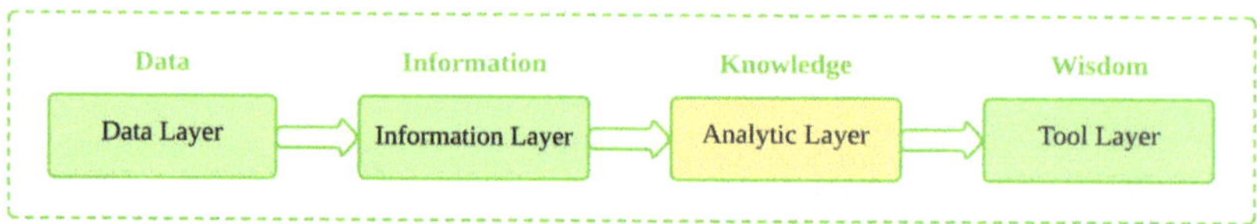

Figure 12. Framework of integration follows the knowledge management model [136].

The four layers mentioned in Figure 11 were connected to a set of data information that belongs to the indoor environmental factors such as noise, temperature, air quality, light, etc. The environment data can be used before merging to other information from different sources in analytic activities. This set-point has been used to analyze the data from the sensors that belong to several facility management. The whole set of data information is explained in Figure 13.

6.4.2. Implementing Service-Oriented Architecture

The future approach relies on implementing the Service-Oriented-Architecture (SOA) whose features include service composition, service discovery, asset wrapping, model-driven implementation, loosely coupled, and platform-independent. SOA combines designing software and other services are combined aiming at providing large application functionalities through a communication protocol. SOA's features are very helpful in reducing the complexities of integrating IoT smart devices in BIM. The current proposals of IoT–BIM integration are not satisfactory in generating a unified architecture system for integrating IoT devices in BIM [138–140]. The goal of any new proposal of IoT–BIM integration should consider a new design where data exchange between SOA, web sites services, and integration methods. The new design should be highlighted by a new representation called representational state transfer (REST) in which only IoT nodes are utilized in the process [135].

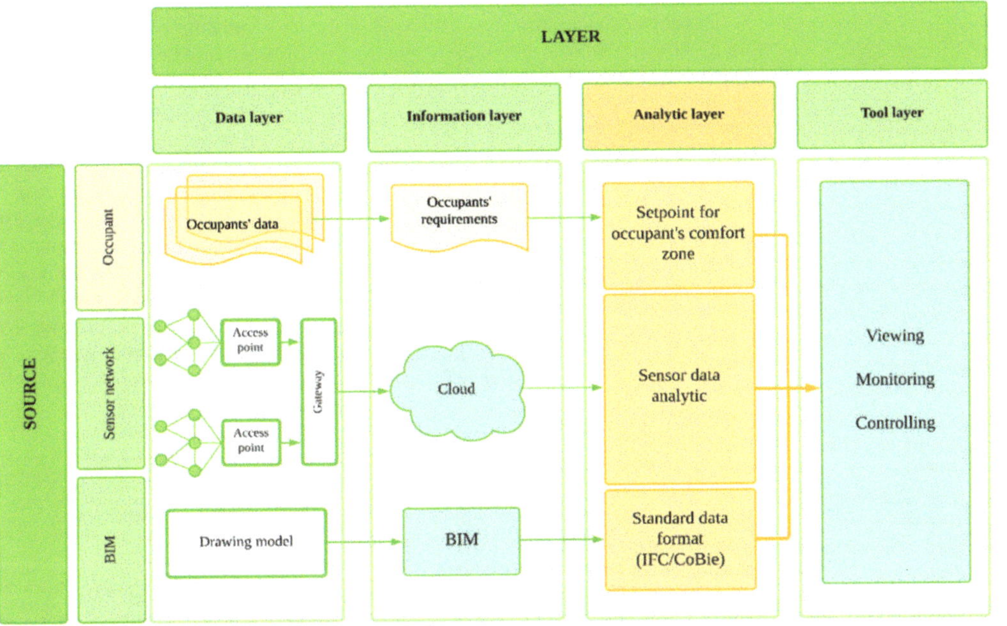

Figure 13. Framework of BIM and sensor integration to improve building performance for occupants' perspective [136].

The first item in developing a new IoT–BIM integration is to reconsider the sematic information of BIM that failed to display elements of indoor conditions. The link of updating static models towards real-time models is necessary to be taken into account. This step results in developing new SOA patterns through which BIM can update the reading of IoT devices. This step requires a system that can do four operations of creating, reading, updating, and deleting (CRUD). The CRUD operations still need heavy investigation. Another point to achieve better IoT–BIM integration is enabling IoT devices to read from models using fusing multi-source information in SOA patterns. Currently, this fusion is carried out by one-way interaction, which develops mimic the human-two-way approach by offering algorithms characterized by human cognitive buildings [141].

Recently, researchers have been focusing on confining the heterogeneous data sources and apply them into different domains to serve intended purposes [77]. Meanwhile, certain issues should be considered such as information consistency, traceability, and archiving [142]. Solving these issues has been proposed by the National Institute of Standards and Technology (NIST) by issuing a framework for Cyber–Physical System (CPS). CPS is a huge system where dimensions such as common language, architecture foundations, and taxonomy are fused to better exchanging ideas and, meanwhile, developing new IoT applications that can be integrated easily with BIM. This approach requires providing comprehended data to handle by the AEC industry. It includes evaluating the query representations within cross-domain data sets, managing collected data for the highest utilization, and assuring a unified information flow during the building lifecycle [143].

In the above approaches of IoT integration, IoT was purely considered as highly related to sensors and actuators. The concept of IoT could be far from these specific issues as new information tools are needed to interconnecting sensors and actuators. This is a high-tech approach that requires highly involving Cloud computations to better connect IoT devices throughout Internet infrastructure [140]. For this reason, Cloud can be adopted in the AEC industry. Currently, most sensors are not connected using Cloud and in this new framework, sensors should be identified and connected in terms of Cloud computing.

The above discussion opens the way to achieve building a system that involves IoT, BIM, and Cloud. Hence, developing generic architecture could be very valuable in building applications that share common requirements and characteristics for the future of BIM.

7. Summary of Previous Empirical Research

Table 2 shows the most recent journal papers that aligned with the contents and the purposes of this current review paper. The first purpose is to integrate data into BIM. It has been proposed integrating cloud data and hyperspectral imaging using high-tech instrumentations equipped with laser and 3D-technique [12]. The study showed that such integration was successfully performed for management facilities. In previous moves, [19] integrated big data in the construction industry to provide new opportunities and future trends. In their attempt, they used a huge number of publications that were collected during the years around 2016. The review concludes that Big Data could have a potential approach to treat the potential pitfalls associated with Big Data adoption in the industry.

Table 2. Summary of previous empirical research.

Specifications	Purpose	Methodology	Main Points	Conclusions
[12] Integration of point cloud data and hyperspectral imaging as a data gathering methodology for refurbishment projects using Building Information Modelling (BIM). Journal of Facilities Management, 17(1), 57–75	Integration of point cloud data and hyperspectral imaging as a data gathering methodology for refurbishment projects using building information modelling (BIM). Journal of Facilities Management.	Laser scanning can be used to collect geometrical and spatial information in the form of a 3D point cloud, and this technique is already used. However, as a point cloud representation does not contain any semantic information or geometrical context, such point cloud data must refer to external sources of data, such as building specification and construction materials, to be in used in BIM.	Hyperspectral imaging techniques can be applied to provide both spectral and spatial information of scenes as a set of high-resolution images. Integrating of a 3D point cloud into hyperspectral images would enable accurate identification and classification of surface materials and would also convert the 3D representation to BIM	This integrated approach was applied in facilities management and construction to improve the efficiency and automation of the data transition from building pathology to BIM. This study integrates laser scanning and hyperspectral imaging. In addition, the study uses a new integration technique which is applied for the first time in the context of buildings.
[19] Big Data in the construction industry: A review of present status, opportunities, and future trends. Advanced engineering informatics, 30(3), 500–521.	Integrate big data in the construction industry to provide new opportunities and future trends.	Related works were reviewed based on publications of the databases of American Association of Civil Engineers (ASCE), Institute of Electrical and Electronics Engineers (IEEE), Association of Computing Machinery (ACM), and Elsevier Science Direct Digital Library.	This paper fills the void and presents a wide-ranging interdisciplinary review of literature of fields such as statistics, data mining and warehousing, machine learning, and Big Data Analytics in the context of the construction industry.	The current state of adoption of Big Data in the construction industry was reviewed. Future potential of Big Data across the multiple domain-specific sub-areas of the construction industry. The review concludes that Big Data could have potential approach to treat the potential pitfalls associated with Big Data adoption in the industry.
[122] Towards a semantic Construction Digital Twin: Directions for future research. Automation in Construction, 114, 103179.	Implementing a semantic Construction Digital Twin: Directions for future research.	Introducing a standardized semantic representation of building components and systems using the Digital Twin conveys. Digital Twin is characterized by socio-technical and process-oriented characterization of the complex artefacts.	The review discusses the multi-faceted applications of BIM during the construction stage and highlights limits and requirements, paving the way to the concept of a Construction Digital Twin.	The study adopted the Digital Twin paradigm in the construction industry sector. Due to this technology, the concept of BIM gained sufficient recognition and momentum to enable a shift from a static, closed information environment to a dynamic.
[30] BIM integrated smart monitoring technique for building fire prevention and disaster relief. Automation in Construction, 84, 14–30.	To integrate smart monitoring technique for building fire prevention and disaster relief BIM.	BIM was used to construct a BIM-based Intelligent Fire Prevention and Disaster Relief System. The methodology uses personal localization, on evacuation/rescue route optimization with Bluetooth-based technology, and on a mobile guidance device to create an intelligent and two-way fire disaster prevention system.	The results of applying the BIM-based system demonstrate that it may effectively provide 3D visualization to support the assessment and planning of fire safety.	The study contributes with providing early detection and alarm responses that is used for efficient evacuation and to facilitate fire rescue and control efforts in order to increase overall building safety and disaster-response capabilities.

Table 2. *Cont.*

Specifications	Purpose	Methodology	Main Points	Conclusions
[50] Building Information Modeling (BIM) for transportation infrastructure–Literature review, applications, challenges, and recommendations. Automation in Construction, 94, 257–281.	Using BIM for improving the transportation infrastructure.	Develop more efficient and cost-effective techniques necessary to repair, advance, and expand the transportation infrastructure.	The results show that the use of BIM for transportation infrastructure has been increasing. More specifically, the research has mainly been focusing on roads, highways, and bridges.	There is a major need for a standard neutral exchange format and schema to promote interoperability. In addition, the continuing collaboration between academia and industry is required to mitigate most challenges and to realize the full potential of BIM for transportation infrastructure.
[117] A framework for integrating BIM and IoT through open standards. Automation in Construction, 95, 35–45.	To create diverse fields including BIM, information system, Automation Systems, and IoT devices for the end users.	The methodology is to integrate the data with IoT sensors and web-based system called Otaniemi3D to integrate BIM and IoT devices through open messaging standards open message interface (O-MI) ad open data format (O-DF) and IFC models.	The paper describes the design criteria, the system architecture, the workflow, and a proof of concept with potential use cases that integrate IoT with the built environment.	The end users and other research groups can benefit from such platforms by either consuming the data in their daily life or using the data for more advance research.
[9] Design and implementation of a novel service management framework for IoT devices in cloud. Journal of Systems and Software, 119, 149–161.	Adopting smart objects to transmit data to the cloud for processing and storage through IoT.	Combining the cloud computing environment with IoT to reduce the transmission and processing cost in the cloud and to provide better services for processing and storing the real time data generated from those IoT devices.	The proposed cloud framework combines IoT and cloud environment to provide services to both IoT and non-IoT users.	A novel framework is designed for the cloud to manage the real time IoT data and scientific non-IoT data. The other part of the framework is cloud, where data storage and process are carried out depending on the user requirement.
[55] Automatic reconstruction of 3D building models from scanned 2D floor plans. Automation in Construction, 63, 48–56.	To significantly improve the systematic use of Information and Communication Technologies (ICT) tools and BIM.	The present article introduces a research work aiming at the development of methods for the generation of 3D building models from 2D plans.	A prototype can extract information from 2D plans and to generate IFC to include the main components of the building: walls, openings, and spaces.	Results are very promising and show that such solutions could be key components of future digital toolkits for renovation design.
[109] Building performance optimization: a hybrid architecture for the integration of contextual information and time-series data. Automation in Construction, 70, 51–61.	To build sematic data for better feasibility of creating adapters between many different software tools.	Presenting a new solution to the semantic data by a hybrid architecture that links data which is retained in its original format. The architecture links existing and efficient relational databases storing time-series data and semantically described building contextual data.	The main contribution of this work is an original RDF syntax structure and ontology to represent existing database schema information, and a new mechanism that automatically prepares data streams for processing by rule-based performance definitions.	The hybrid architecture avoids the duplication of time-series data and overcomes some of the differences found in database schemas and database platforms.
[144] An IoT-based autonomous system for workers' safety in construction sites with real-time alarming, monitoring, and positioning strategies. Automation in Construction, 88, 73–86.	To protect construction workers and prevent accidents in such sites.	The design of the wearable device includes a set of components which are a radio transceiver (transmitter/receiver), a wake-up sensor, an alarm actuator, and a General Packet Radio Service (GPRS) module.	The heterogeneous components of this architecture are seamlessly integrated into a middleware backend online server. The wearable device has a power saving scheme with a current consumption as low as 0.5 µA at 3 V.	Presenting an implementation of wireless nodes that are powered by light energy using photovoltaic cells. These nodes adopt energy management and storage schemes for continuous operation for indoor and outdoor environments.
[119] Top 10 technologies for indoor positioning on construction sites. Automation in Construction, 118, 103309.	Demonstrating indoor positioning enables five significant applications that considerably enhance work efficiency and safety on construction sites	Indoor positioning systems can be viewed as a combination of (1) creating corresponding algorithms, (2) indoor positioning technologies, and (3) indoor positioning hardware equipment.	Full analysis for challenges in applying six indoor positioning systems on construction sites. The system was to include technologies and principles.	A promising trends of indoor positioning development for indoor positioning hybridization was created using game theory positioning, and integration with BIM model.
[37] Metadata Models and Methods for Smart Buildings (Doctoral dissertation, UC San Diego).	To use an effective sensing data and Heating, ventilation, and air conditioning (HVAC), security, lighting and sensing subsystems.	We envision building systems to exchange data across subsystems as well as across various building services in a programming framework. Such information exchange is mediated by timely sensor information.	A programming framework comprised of machine learning algorithms was developed relying on a standard information model for unified and secure application deployment	Demonstrating new devices such as thermostat called Genie, an energy dashboard, and a metadata models for building portable applications for smart buildings in this dissertation, we continue to pursue building a community of system builders for the smart building environments.

Another attempt to integrate the data was performed by [123] in which semantic Construction Digital Twin was used in BIM which was apparently gained recognition. A smart monitoring technique for building fire prevention and disaster relief was also integrated into BIM hoping to reduce the fatalities and damages [30]. The results of this integration have shown a serious contribution to public safety. In another move of IoT integration in BIM, the role of BIM was improved in the transportation infrastructure [50]. The study aimed at developing more efficient and cost-effective techniques necessary to repair, advance, and expand the transportation infrastructure.

Moreover, [117] created diverse fields including BIM, information systems, Automation Systems, and IoT devices for the end-users. The end-users and other research groups benefited from such platforms by either consuming the data in their daily life or using the data for more advanced research.

Reference [9] designed and implemented a novel service management framework for IoT devices in the cloud computing system. The combination of the cloud computing environment and IoT has resulted in reducing the transmission and processing cost in the cloud and providing better services for processing and storing the real-time data generated from those IoT devices. Reference [55] significantly improved the systematic use of Information and Communication Technologies (ICT) tools and (BIM). The results were very promising and showed that such solutions could be key components of future digital toolkits for renovation design. Reference [109] used hybrid architecture for the integration of contextual information and time-series data.

This was another trial for implementing a new solution to the semantic data which was retained in its original format. This approach requires efficient relational databases storing time-series data and semantically described building contextual data. The most important part in this hybridization was to avoid the duplication of time-series data and to overcome some of the differences found in database schemas and database platforms.

Moreover, [140] created an IoT-based autonomous system for workers' safety in construction sites with real-time alarming, monitoring, and positioning strategies. The study extended the previous study of fire safety as noted earlier by [30].

In a very recent study, [119] recognized the top 10 technologies for indoor positioning on construction sites to demonstrate indoor positioning to enabling five significant applications that considerably enhance work efficiency and safety on construction sites. The last study performed by [37] in which metadata models and methods for Smart Buildings were proposed for using effective sensing data and (HVAC), security, lighting, and sensing subsystems. The study demonstrated new devices such as a thermostat called Genie, an energy dashboard, and a metadata models for building portable applications for smart buildings in this dissertation, we continue to pursue building a community of system builders for the smart building environments.

8. Contribution

The current review stressed the Semantic Web and relational databases to show a successful attempt to store cross-domain data. The procedure to achieve this goal was performed by three steps compromising to represent contextual information (sensor information and other soft building information) in RDF format, retaining data gathered by time-series data in the relational database, and mapping contextual information using sensor ID described in RDF.

The web and relational information approach are the most important procedure that could serve interlinking between various data sources. As such, a model was developed showing the effectiveness of storing time-series data in the relational model. The outcome shows the effectiveness of the query language resulted from integrating SPARQL and SQL. These developments were highly considered as the most promising methods to facilitate IoT deployment first and then to integrate the information in BIM and creating platforms and formats that were suitable for BIM–IoT integration. The other contribution of this study is to utilize standardized data formats and query language into specific domain data sources to extend project scope.

9. Conclusions

The information integration of smart industries is an important step toward better understanding the construction of the renovation of existing buildings. Despite developing smart technologies such as BIM and IoT, it was found that a single technology may face challenges that result in integrating two or technologies to deal with evolving challenges. It has been reported that BIM implementation requires the live data in addition to other physical data. For these consequences, models for BIM implementation should be developed and applied. The most important step is to the Service-Oriented-Architecture (SOA) whose features include service composition, service discovery, asset wrapping, model-driven implementation, loosely coupled, and platform-independent. SOA combines designing software and other services are combined aiming at providing large application functionalities through a communication protocol. The first item in developing a new IoT–BIM integration is to reconsider the sematic information of BIM that was failed to display elements of indoor conditions. The second challenge is about the size of data stored in the system and the suitability of users to utilize this data. Several procedures were proposed to create open storage. BIM works very well for new developments; however, for the existing building, the role of BIM may become much harder. The current study contributes valuable information gathered from a long list of publications which are mainly during the last three years. The gathered information could benefit AEC, construction operation, monitoring, health and safety, and FM. It is also, this study sheds the light on the challenges and limitations of high-tech tools such as cameras and sensors. The goal of this study is to provide the best methodology to create a suitable database. The database created via BIM–IoT integration utilizes the Application Programming Interface (API) and relational data by creating new queries, language, semantic web technology, and hybridization. This study concludes that the possibility of prominent future research through which solving interoperability issues and cloud computing are heavily used. The digital revolution is the most important support of much potential BIM and IoT individually and under integration. This study also helps researchers to transform the traditional construction engineering to many advance concepts where visualization and industrial foundation cases are practiced. Updating static models towards real-time models is very important step which could result in developing new SOA patterns through which BIM can enable to update of the reading of IoT devices. Currently, this fusion is carried out by one-way interaction which could be developed to mimic the human-two-way approach by offering algorithms characterized by human cognitive buildings. Meanwhile, certain issues should be considered such as information consistency, traceability, and archiving. This is a high-tech approach that requires highly involving Cloud computations to better connect IoT devices throughout Internet infrastructure.

Author Contributions: Conceptualization, A.B.A.A. and N.A.H.; methodology, A.B.A.A.; writing—original draft preparation, A.B.A.A.; writing—review and editing, A.B.A.A.; supervision and reviewed the article, N.A.H.; Advice and support, A.H.A. and T.H.L. All authors have read and agreed to the published version of the manuscript.

Funding: This work was supported and partial funding by Research Management Centre, University Putra Malaysia.

Institutional Review Board Statement: Not Applicable.

Informed Consent Statement: Not Applicable.

Data Availability Statement: Not Applicable.

Acknowledgments: The authors would like to thank the Department of Civil Engineering, Faculty of Engineering, University Putra Malaysia, Serdang, Malaysia, for supporting this research paper.

Conflicts of Interest: The authors declare no conflict of interest.

Abbreviations

2D/3D	Two/Three-Dimensional
3V's	Volume, Variety, Velocity
ACM	Association for Computing Machinery
AEC	Architecture, Engineering and Construction
AECO	Architecture, Engineering, Construction, and Operation
AI	Artificial Intelligence
API	Application Programming Interface
ASCE	American Society of Civil Engineers
BDA	Big Data Analytics
BDE	Big Data Engineering
BHIMM	Heritage Information Modelling and Management
BIM	Building Information Modelling
BIMQL	BIM Query Language
CAD	computer-aided design
CPS	Cyber–Physical System
CRUD	Create, Read, Update and Delete.
DAG	Directed Acrylic Graphs
DG	Digital Photogrammetry
DM	Data Mining
DOC/XLS/PPT	Microsoft Office Format
DWG	Drawing File Format.
DXF	Drawing Exchange Format
EMR	Elastic MapReduce
FM	Facility Management
GPR	Ground Penetration Radar
GPRS	General Packet Radio Service
GUI	Graphic User Interface
Hadoop	High Availability Distributed Object-Oriented Platform
HBase™	Hadoop Base.
HBIM	Historic Building Information Modelling
HVAC	Heating, Ventilation, and Air Conditioning
ICT	Information and Communication Technologies
IFC	Industry Foundation Classes
ifcOWL	Industry Foundation Classes OWL
ifcXML	Industry Foundation Classes XML
IO	Input/output data
IoT	Internet of Things
IRT	Infrared thermography images
JPEG	Image Format
KM Analysis	Knowledge Management Analysis
LC	Life Cycle
LIFE	Lean and Injury-Free
LoD	Level of Detail
MR	Mappers and Reducers
NDT	Non-Destructive Testing
NIST	National Institute of Standards and Technology
NLQ	Natural Language Query
O&M	Operation and Maintenance
OntoFM	Ontology Facility Management
OWL	Web Ontology Language
RDF	Resource Description Format
REST	Representational State Transfer
RFIDs	Radio Frequency Identifications
RGB	Red, Green, Blue
RM)/MPG	Video Format

RVT (Revit)	Software for BIM
SAREF	Smart Appliances Reference ontology
SOA	Service-Oriented-Architecture
SPARQL	Query Language and Protocol.
SQL	Structured Query Language
SSN	Semantic Sensor Network
TLS	Terrestrial Laser Scanner
UAS	Unmanned Aerial Systems
XQuery	XML Query)

References

1. Xu, X.; Mumford, T.; Zou, P.X. Life-cycle building information modelling (BIM) engaged framework for improving building energy performance. *Energy Build.* **2021**, *231*, 110496. [CrossRef]
2. Wang, C.; Cho, Y.K.; Kim, C. Automatic BIM component extraction from point clouds of existing buildings for sustainability applications. *Autom. Constr.* **2015**, *56*, 1–13. [CrossRef]
3. Ansah, M.K.; Chen, X.; Yang, H.; Lu, L.; Lam, P.T. A review and outlook for integrated BIM application in green building assessment. *Sustain. Cities Soc.* **2019**, *48*, 101576. [CrossRef]
4. Lu, Y.; Wu, Z.; Chang, R.; Li, Y. Building Information Modeling (BIM) for green buildings: A critical review and future directions. *Autom. Constr.* **2017**, *83*, 134–148. [CrossRef]
5. Gao, J.; Zhong, X.; Cai, W.; Ren, H.; Huo, T.; Wang, X.; Mi, Z. Dilution effect of the building area on energy intensity in urban residential buildings. *Nat. Commun.* **2019**, *10*, 4944. [CrossRef]
6. Pourzolfaghar, Z.; McDonnell, P.; Helfert, M. Barriers to benefit from integration of building information with live data from IOT devices during the facility management phase. In Proceedings of the CITA BIM Gathering 2015, Dublin, Ireland, 23–24 November 2017.
7. Gubbi, J.; Buyya, R.; Marusic, S.; Palaniswami, M. Internet of Things (IoT): A vision, architectural elements, and future directions. *Future Gener. Comput. Syst.* **2013**, *29*, 1645–1660. [CrossRef]
8. Čolaković, A.; Hadžialić, M. Internet of Things (IoT): A review of enabling technologies, challenges, and open research issues. *Comput. Netw.* **2018**, *144*, 17–39. [CrossRef]
9. Dehury, C.K.; Sahoo, P.K. Design and implementation of a novel service management framework for IoT devices in cloud. *J. Syst. Softw.* **2016**, *119*, 149–161. [CrossRef]
10. Tang, S.; Shelden, D.R.; Eastman, C.M.; Pishdad-Bozorgi, P.; Gao, X. A review of building information modeling (BIM) and the internet of things (IoT) devices integration: Present status and future trends. *Autom. Constr.* **2019**, *101*, 127–139. [CrossRef]
11. Teizer, J.; Wolf, M.; Golovina, O.; Perschewski, M.; Propach, M.; Neges, M.; König, M. Internet of Things (IoT) for integrating environmental and localization data in Building Information Modeling (BIM). In *ISARC Proceedings of the International Symposium on Automation and Robotics in Construction (Vol. 34)*; Vilnius Gediminas Technical University: Vilnius, Lithuania, 2017.
12. Amano, K.; Lou, E.C.; Edwards, R. Integration of point cloud data and hyperspectral imaging as a data gathering methodology for refurbishment projects using Building Information Modelling (BIM). *J. Facil. Manag.* **2019**, *17*, 57–75. [CrossRef]
13. Ozturk, G.B. Interoperability in building information modeling for AECO/FM industry. *Autom. Constr.* **2020**, *113*, 103122. [CrossRef]
14. Akcamete, A.; Akinci, B.; Garrett, J.H., Jr. Motivation for computational support for updating building information models (BIMs). In *Computing in Civil Engineering*; ASCE: Reston, VA, USA, 2009; pp. 523–532.
15. Arayici, Y. Towards Building Information Modelling for Existing Structures. Structural Survey. Available online: http://usir.salford.ac.uk/id/eprint/12473/ (accessed on 27 February 2021).
16. Armesto, J.; Lubowiecka, I.; Ordóñez, C.; Rial, F.I. FEM modeling of structures based on close range digital photogrammetry. *Autom. Constr.* **2009**, *18*, 559–569. [CrossRef]
17. Klein, L.; Li, N.; Becerik-Gerber, B. Imaged-based verification of as-built documentation of operational buildings. *Autom. Constr.* **2012**, *21*, 161–171. [CrossRef]
18. ISO 22263:2008. *Organization of Information About Construction Works—Framework for Management of Project Information, International Standard*; ISO: Geneva, Switzerland, 2008.
19. Bilal, M.; Oyedele, L.O.; Qadir, J.; Munir, K.; Ajayi, S.O.; Akinade, O.O.; Pasha, M. Big Data in the construction industry: A review of present status, opportunities, and future trends. *Adv. Eng. Inform.* **2016**, *30*, 500–521. [CrossRef]
20. Zhao, X. A scientometric review of global BIM research: Analysis and visualization. *Autom. Constr.* **2017**, *80*, 37–47. [CrossRef]
21. Ding, K.; Shi, H.; Hui, J.; Liu, Y.; Zhu, B.; Zhang, F.; Cao, W. Smart steel bridge construction enabled by BIM and Internet of Things in industry 4.0: A framework. In Proceedings of the 2018 IEEE 15th International Conference on Networking, Sensing and Control (ICNSC), Zhuhai, China, 27–29 March 2018; IEEE: New York, NY, USA, 2018; pp. 1–5.
22. Fang, Y.; Cho, Y.K.; Durso, F.; Seo, J. Assessment of operator's situation awareness for smart operation of mobile cranes. *Autom. Constr.* **2018**, *85*, 65–75. [CrossRef]
23. Rahimian, F.P.; Seyedzadeh, S.; Oliver, S.; Rodriguez, S.; Dawood, N. On-demand monitoring of construction projects through a game-like hybrid application of BIM and machine learning. *Autom. Constr.* **2020**, *110*, 103012. [CrossRef]

24. Zhong, R.Y.; Peng, Y.; Xue, F.; Fang, J.; Zou, W.; Luo, H.; Ng, S.T.; Lu, W.; Shen, G.Q.; Huang, G.Q. Prefabricated construction enabled by the Internet-of-Things. *Autom. Constr.* **2017**, *76*, 59–70. [CrossRef]
25. Xu, G.; Li, M.; Chen, C.H.; Wei, Y. Cloud asset-enabled integrated IoT platform for lean prefabricated construction. *Autom. Constr.* **2018**, *93*, 123–134. [CrossRef]
26. Li, C.Z.; Xue, F.; Li, X.; Hong, J.; Shen, G.Q. An Internet of Things-enabled BIM platform for on-site assembly services in prefabricated construction. *Autom. Constr.* **2018**, *89*, 146–161. [CrossRef]
27. Mostafa, S.; Kim, K.P.; Tam, V.W.; Rahnamayiezekavat, P. Exploring the status, benefits, barriers and opportunities of using BIM for advancing prefabrication practice. *Int. J. Constr. Manag.* **2020**, *20*, 146–156. [CrossRef]
28. Pärn, E.A.; Edwards, D.J. Conceptualising the FinDD API plug-in: A study of BIM-FM integration. *Autom. Constr.* **2017**, *80*, 11–21. [CrossRef]
29. Habibi, S. Micro-climatization and real-time digitalization effects on energy efficiency based on user behavior. *Build. Environ.* **2017**, *114*, 410–428. [CrossRef]
30. Cheng, M.Y.; Chiu, K.C.; Hsieh, Y.M.; Yang, I.T.; Chou, J.S.; Wu, Y.W. BIM integrated smart monitoring technique for building fire prevention and disaster relief. *Autom. Constr.* **2017**, *84*, 14–30. [CrossRef]
31. Dixit, M.K.; Venkatraj, V.; Ostadalimakhmalbaf, M.; Pariafsai, F.; Lavy, S. Integration of facility management and building information modeling (BIM). *Facilities* **2019**, *37*, 455–483. [CrossRef]
32. Kim, K.; Cho, Y.; Zhang, S. Integrating work sequences and temporary structures into safety planning: Automated scaffolding-related safety hazard identification and prevention in BIM. *Autom. Constr.* **2016**, *70*, 128–142. [CrossRef]
33. Eastman, C.M.; Teicholz, P.; Sacks, R.; Liston, K. *Handbook BIM: A Guide to Building Information Modeling for Owners, Managers, Architects, Engineers, Contractors, and Fabricators*; Wiley: Hoboken, NJ, USA, 2008.
34. Nawari, N.O.; Ravindran, S. Blockchain and building information modeling (BIM): Review and applications in post-disaster recovery. *Buildings* **2019**, *9*, 149. [CrossRef]
35. Shepherd, D. *The BIM Management Handbook*; Routledge: London, UK, 2019.
36. Bruno, N.; Roncella, R. HBIM for conservation: A new proposal for information modeling. *Remote Sens.* **2019**, *11*, 1751. [CrossRef]
37. Koh, J.B. Metadata Models and Methods for Smart Buildings. Ph.D. Thesis, University of California San Diego, San Diego, CA, USA, 2020.
38. Murphy, M.; McGovern, E.; Pavia, S. Historic building information modelling (HBIM). *Struct. Surv.* **2009**, *27*, 311–327. [CrossRef]
39. Bruno, S.; De Fino, M.; Fatiguso, F. Historic Building Information Modelling: Performance assessment for diagnosis-aided information modelling and management. *Autom. Constr.* **2018**, *86*, 256–276. [CrossRef]
40. Ciribini, A.L.C.; Ventura, S.M.; Paneroni, M. BIM methodology as an integrated approach to heritage conservation management. *Wit Trans. Built Environ.* **2015**, *149*, 265–276.
41. Ilter, D.; Ergen, E. BIM for building refurbishment and maintenance: Current status and research directions. *Struct. Surv.* **2015**, *33*, 228–256. [CrossRef]
42. Volk, R.; Stengel, J.; Schultmann, F. Building Information Modeling (BIM) for existing buildings—Literature review and future needs. *Autom. Constr.* **2014**, *38*, 109–127. [CrossRef]
43. Oreni, D.; Brumana, R.; Georgopoulos, A.; Cuca, B. HBIM for conservation and management of built heritage: Towards a library of vaults and wooden bean floors. *ISPRS* **2013**, *5*, W1. [CrossRef]
44. Sun, Z.; Xie, J.; Zhang, Y.; Cao, Y. As-Built BIM for a fifteenth-century Chinese brick structure at various LoDs. *ISPRS Int. J. Geo-Inf.* **2019**, *8*, 577. [CrossRef]
45. Fai, S.; Graham, K.; Duckworth, T.; Wood, N.; Attar, R. Building information modelling and heritage documentation. In Proceedings of the 23rd International Symposium, International Scientific Committee for Documentation of Cultural Heritage (CIPA), Prague, Czech Republic, 11–16 September 2011; pp. 12–16.
46. McArthur, J.J. A building information management (BIM) framework and supporting case study for existing building operations, maintenance and sustainability. *Procedia Eng.* **2015**, *118*, 1104–1111. [CrossRef]
47. Chiesa, G. ICT, Data and Design Issues. In *Technological Paradigms and Digital Eras*; Springer: Cham, Switzerland, 2020; pp. 1–38.
48. Cruz, I.F.; Xiao, H. Ontology driven data integration in heterogeneous networks. In *Complex Systems in Knowledge-Based Environments: Theory, Models and Applications*; Springer: Berlin/Heidelberg, Germany, 2009; pp. 75–98.
49. Hu, Z.Z.; Tian, P.L.; Li, S.W.; Zhang, J.P. BIM-based integrated delivery technologies for intelligent MEP management in the operation and maintenance phase. *Adv. Eng. Softw.* **2018**, *115*, 1–16. [CrossRef]
50. Costin, A.; Adibfar, A.; Hu, H.; Chen, S.S. Building Information Modeling (BIM) for transportation infrastructure–Literature review, applications, challenges, and recommendations. *Autom. Constr.* **2018**, *94*, 257–281. [CrossRef]
51. Zhai, Y.; Chen, K.; Zhou, J.X.; Cao, J.; Lyu, Z.; Jin, X.; Shen, G.Q.; Lu, W.; Huang, G.Q. An Internet of Things-enabled BIM platform for modular integrated construction: A case study in Hong Kong. *Adv. Eng. Inform.* **2019**, *42*, 100997. [CrossRef]
52. Hossain, M.A.; Yeoh, J.K.W. BIM for Existing Buildings: Potential Opportunities and Barriers. In *IOP Conference Series: Materials Science and Engineering*; IOP Publishing: Bristol, UK, 2018; Volume 371, p. 012051.
53. Motawa, I.; Almarshad, A. A knowledge-based BIM system for building maintenance. *Autom. Constr.* **2013**, *29*, 173–182. [CrossRef]

54. Ghaffarianhoseini, A.; Tookey, J.; Ghaffarianhoseini, A.; Naismith, N.; Azhar, S.; Efimova, O.; Raahemifar, K. Building Information Modelling (BIM) uptake: Clear benefits, understanding its implementation, risks and challenges. *Renew. Sustain. Energy Rev.* **2017**, *75*, 1046–1053. [CrossRef]
55. Gimenez, L.; Robert, S.; Suard, F.; Zreik, K. Automatic reconstruction of 3D building models from scanned 2D floor plans. *Autom. Constr.* **2016**, *63*, 48–56. [CrossRef]
56. NBS. *National BIM Report 2015*; Royal Institute of British Architects: London, UK, 2015.
57. Lu, Q.; Lee, S. Image-based technologies for constructing as-is building information models for existing buildings. *J. Comput. Civ. Eng.* **2017**, *31*, 04017005. [CrossRef]
58. Laefer, D.F.; Truong-Hong, L. Toward automatic generation of 3D steel structures for building information modelling. *Autom. Constr.* **2017**, *74*, 66–77. [CrossRef]
59. Lucas, J.D. Managing the Facility with Lifecycle Information. *J. Curr. Issues Media Telecommun.* **2015**, *7*, 13–36.
60. Teicholz, E. Bridging the AEC technology gap. *IFMA Facil. Manag. J.* **2004**, *587*, 588.
61. IFMA. *Exploring the Current Trends and Future Outlook for Facility Management*; Facility Management Forecast 2011; The International Facility Management Association: Houston, TX, USA, 2011; ISBN 1-883176-84-0.
62. Saberi, S.; Kouhizadeh, M.; Sarkis, J.; Shen, L. Blockchain technology and its relationships to sustainable supply chain management. *Int. J. Prod. Res.* **2019**, *57*, 2117–2135. [CrossRef]
63. Ma, Z.; Cooper, P.; Daly, D.; Ledo, L. Existing building retrofits: Methodology and state-of-the-art. *Energy Build.* **2012**, *55*, 889–902. [CrossRef]
64. Shaikh, P.H.; Shaikh, F.; Sahito, A.A.; Uqaili, M.A.; Umrani, Z. An Overview of the Challenges for Cost-Effective and Energy-Efficient Retrofits of the Existing Building Stock. In *Cost-Effective Energy Efficient Building Retrofitting*; Woodhead Publishing: Swaston, UK, 2017; pp. 257–278.
65. Asadi, E.; Da Silva, M.G.; Antunes, C.H.; Dias, L. Multi-objective optimization for building retrofit strategies: A model and an application. *Energy Build.* **2012**, *44*, 81–87. [CrossRef]
66. Huang, Z.; Ge, J.; Zhao, K.; Shen, J. Post-evaluation of energy consumption of the green retrofit building. *Energy Procedia* **2019**, *158*, 3608–3613. [CrossRef]
67. Salvalai, G.; Sesana, M.M.; Iannaccone, G. Deep renovation of multi-storey multi-owner existing residential buildings: A pilot case study in Italy. *Energy Build.* **2017**, *148*, 23–36. [CrossRef]
68. Xiaonuan, S.; SiuYu, S.L. Existing buildings' operation and maintenance: Renovation project of Chow Yei Ching Building at the University of Hong Kong. *Int. J. Low-Carbon Technol.* **2014**, *10*, 393–404. [CrossRef]
69. Pärn, E.A.; Edwards, D.J.; Sing, M.C. The building information modelling trajectory in facilities management: A review. *Autom. Constr.* **2017**, *75*, 45–55. [CrossRef]
70. Kassem, M.; Kelly, G.; Dawood, N.; Serginson, M.; Lockley, S. BIM in facilities management applications: A case study of a large university complex. *Built Environ. Proj. Asset. Manag.* **2015**, *5*, 261–277. [CrossRef]
71. Bortoluzzi, B.; Efremov, I.; Medina, C.; Sobieraj, D.; McArthur, J.J. Automating the creation of building information models for existing buildings. *Autom. Constr.* **2019**, *105*, 102838. [CrossRef]
72. Dixit, M.K.; Venkatraj, V. Integrating facility management functions in building information modeling (BIM): A review of key issues and challenges. In Proceedings of the International Research Conference (IRC) 2017: Shaping Tomorrow's Built Environment Conference Proceedings, Salford, UK, 11–12 September 2017; pp. 597–608.
73. Carli, R.; Dotoli, M.; Pellegrino, R.; Ranieri, L. Using multi-objective optimization for the integrated energy efficiency improvement of a smart city public buildings' portfolio. In Proceedings of the 2015 IEEE International Conference on Automation Science and Engineering (CASE), Gothenburg, Sweden, 24–28 August 2015; pp. 21–26.
74. Peng, Y.; Lin, J.R.; Zhang, J.P.; Hu, Z.Z. A hybrid data mining approach on BIM-based building operation and maintenance. *Build. Environ.* **2017**, *126*, 483–495. [CrossRef]
75. Quattrini, R.; Pierdicca, R.; Morbidoni, C. Knowledge-based data enrichment for HBIM: Exploring high-quality models using the semantic-web. *J. Cult. Herit.* **2017**, *28*, 129–139. [CrossRef]
76. Wanner, E. *On Remembering, Forgetting, and Understanding Sentences: A Study of the Deep Structure Hypothesis*; Walter de Gruyter GmbH Co KG: Berlin, Germany, 2019; Volume 170.
77. Pourzolfaghar, Z.; Helfert, M. Integration of buildings information with live data from IoT devices. In *Connected Environments for the Internet of Things*; Springer: Cham, Switzerland, 2017; pp. 169–185.
78. Dexeus, C.R. The deepening effects of the digital revolution. In *The Future of Tourism*; Springer: Cham, Switzerland, 2019; pp. 43–69.
79. Eadie, R.; Browne, M.; Odeyinka, H.; McKeown, C.; McNiff, S. BIM implementation throughout the UK construction project lifecycle: An analysis. *Autom. Constr.* **2013**, *36*, 145–151. [CrossRef]
80. Jiao, Y.; Zhang, S.; Li, Y.; Wang, Y.; Yang, B.; Wang, L. An augmented Mapreduce framework for building information modeling applications. In Proceedings of the 2014 IEEE 18th International Conference on Computer Supported Cooperative Work in Design (CSCWD), Hsinchu, Taiwan, 21–23 May 2014; IEEE: New York, NY, USA, 2014; pp. 283–288.
81. Lin, J.R.; Hu, Z.Z.; Zhang, J.P.; Yu, F.Q. A Natural-Language-Based Approach to Intelligent Data Retrieval and Representation for Cloud BIM. *Comput. Aided Civ. Infrastruct. Eng.* **2016**, *31*, 18–33. [CrossRef]

82. Zheng, R.; Jiang, J.; Hao, X.; Ren, W.; Xiong, F.; Ren, Y. bcBIM: A blockchain-based big data model for BIM modification audit and provenance in mobile cloud. *Math. Probl. Eng.* **2019**, 2019. [CrossRef]
83. Hofmann, E. Big data and supply chain decisions: The impact of volume, variety and velocity properties on the bullwhip effect. *Int. J. Prod. Res.* **2017**, *55*, 5108–5126. [CrossRef]
84. Provost, F.; Fawcett, T. *Data Science for Business: What You Need to Know about Data Mining and Data-Analytic Thinking*; O'Reilly Media, Inc.: Newton, MA, USA, 2013.
85. Ray, P.D. Pervasive, Domain and Situational-Aware, Adaptive, Automated, and Coordinated Big Data Analysis, Contextual Learning and Predictive Control of Business and Operational Risks and Security. U.S. Patent No. 10,210,470, 19 February 2019.
86. Matsunaga, F.T.; Brancher, J.D.; Busto, R.M. Data mining techniques and tasks for multidisciplinary applications: A systematic review. *ReABTIC* **2015**, 1. [CrossRef]
87. Dean, J.; Ghemawat, S. MapReduce: Simplified data processing on large clusters. *Commun. ACM* **2008**, *51*, 107–113. [CrossRef]
88. Wang, Q.; Lee, B.; Murray, N.; Qiao, Y. MR-IoT: An information centric MapReduce framework for IoT. In Proceedings of the 2018 15th IEEE Annual Consumer Communications & Networking Conference (CCNC), Las Vegas, NV, USA, 12–15 January 2018; pp. 1–6.
89. Qadir, J.; Ahad, N.; Mushtaq, E.; Bilal, M. SDNs, clouds, and big data: New opportunities. In Proceedings of the 2014 12th International Conference on Frontiers of Information Technology, Islamabad, Pakistan, 17–19 December 2014; pp. 28–33.
90. Agneeswaran, V.S. *Big Data Analytics beyond Hadoop: Real-Time Applications with Storm, Spark, and More Hadoop Alternatives*; FT Press: Upper Saddle River, NJ, USA, 2014.
91. Watson, H.J. Update tutorial: Big Data analytics: Concepts, technology, and applications. *Commun. Assoc. Inf. Syst.* **2019**, *44*, 365–379.
92. Lu, J.; Chen, Y.; Herodotou, H.; Babu, S. Speedup your analytics: Automatic parameter tuning for databases and big data systems. *Proc. Vldb Endow.* **2019**, *12*, 1970–1973. [CrossRef]
93. Chen, H.; Chang, P.; Hu, Z.; Fu, H.; Yan, L. A Spark-based Ant Lion Algorithm for Parameters Optimization of Random Forest in Credit Classifi.cation. In Proceedings of the 2019 IEEE 3rd Information Technology, Networking, Electronic and Automation Control Conference (ITNEC), Chengdu, China, 15–17 March 2019; pp. 992–996.
94. Singh, D.; Reddy, C.K. A survey on platforms for big data analytics. *J. Big Data* **2015**, *2*, 8. [CrossRef] [PubMed]
95. Karau, H.; Konwinski, A.; Wendell, P.; Zaharia, M. *Learning Spark: Lightning-Fast Big Data Analysis*; O'Reilly Media, Inc.: Newton, MA, USA, 2015.
96. Garofalo, M.; Botta, A.; Ventre, G. Astrophysics and big data: Challenges, methods, and tools. *Proc. Int. Astron. Union* **2016**, *12*, 345–348. [CrossRef]
97. Corry, E.; O'Donnell, J.; Curry, E.; Coakley, D.; Pauwels, P.; Keane, M. Using semantic web technologies to access soft AEC data. *Adv. Eng. Inform.* **2014**, *28*, 370–380. [CrossRef]
98. Kazmi, A.H.; O'grady, M.J.; Delaney, D.T.; Ruzzelli, A.G.; O'hare, G.M. A review of wireless-sensor-network-enabled building energy management systems. *ACM Trans. Sens. Netw. (TOSN)* **2014**, *10*, 66. [CrossRef]
99. Woo, J.H.; Peterson, M.A.; Gleason, B. Developing a virtual campus model in an interactive game-engine environment for building energy benchmarking. *J. Comput. Civ. Eng.* **2016**, *30*, C4016005. [CrossRef]
100. Lingappa, P.R. Master Applet for Secure Remote Payment Processing. U.S. Patent No. 10,592,899, 17 March 2020.
101. Abu-Tair, M.; Djahel, S.; Perry, P.; Scotney, B.; Zia, U.; Carracedo, J.M.; Sajjad, A. Towards Secure and Privacy-Preserving IoT Enabled Smart Home: Architecture and Experimental Study. *Sensors* **2020**, *20*, 6131. [CrossRef] [PubMed]
102. Motamedi, A.; Hammad, A.; Asen, Y. Knowledge-assisted BIM-based visual analytics for failure root cause detection in facilities management. *Autom. Constr.* **2014**, *43*, 73–83. [CrossRef]
103. Singh, G.; Solanki, A. An algorithm to transform natural language into SQL queries for relational databases. *Selforganizology* **2016**, *3*, 100–116.
104. Mazairac, W.; Beetz, J. BIMQL–An open query language for building information models. *Adv. Eng. Inform.* **2013**, *27*, 444–456. [CrossRef]
105. Schreiber, G.; Akkermans, H.; Anjewierden, A.; Shadbolt, N.R.; de Hoog, R.; Van de Velde, W.; Wielinga, B.J. *Knowledge Engineering and Management: The CommonKADS Methodology*; MIT Press: Cambridge, MA, USA, 2000.
106. Chen, L.; Shadbolt, N.R.; Goble, C.A. A semantic web-based approach to knowledge management for grid applications. *IEEE Trans. Knowl. Data Eng.* **2006**, *19*, 283–296. [CrossRef]
107. Dibley, M.; Li, H.; Rezgui, Y.; Miles, J. An ontology framework for intelligent sensor-based building monitoring. *Autom. Constr.* **2012**, *28*, 1–14. [CrossRef]
108. Curry, E.; O'Donnell, J.; Corry, E.; Hasan, S.; Keane, M.; O'Riain, S. Linking building data in the cloud: Integrating cross-domain building data using linked data. *Adv. Eng. Inform.* **2013**, *27*, 206–219. [CrossRef]
109. Hu, S.; Corry, E.; Curry, E.; Turner, W.J.; O'Donnell, J. Building performance optimisation: A hybrid architecture for the integration of contextual information and time-series data. *Autom. Constr.* **2016**, *70*, 51–61. [CrossRef]
110. Fedushko, S.; Ustyianovych, T. Operational Intelligence Software Concepts for Continuous Healthcare Monitoring and Consolidated Data Storage Ecosystem. In *International Conference on Computer Science, Engineering and Education Applications*; Springer: Cham, Switzerland, 2020; pp. 545–557.

111. Fedushko, S.; Ustyianovych, T.; Gregus, M. Real-time high-load infrastructure transaction status output prediction using operational intelligence and big data technologies. *Electronics* **2020**, *9*, 668. [CrossRef]
112. Ogasawara, G.H.; Tso, M.M. Hybrid Data Management System and Method for Managing Large, Varying Datasets. U.S. Patent No. 9,396,290, 19 July 2016.
113. McGlinn, K.; Yuce, B.; Wicaksono, H.; Howell, S.; Rezgui, Y. Usability evaluation of a web-based tool for supporting holistic building energy management. *Autom. Constr.* **2017**, *84*, 154–165. [CrossRef]
114. Li, J.; Kassem, M.; Ciribini, A.L.C.; Bolpagni, M. A proposed approach integrating DLT, BIM, IOT and smart contracts: Demonstration using a simulated installation task. In *International Conference on Smart Infrastructure and Construction 2019 (ICSIC) Driving Data-Informed Decision-Making*; ICE Publishing: London, UK, 2019; pp. 275–282.
115. Sacks, R.; Eastman, C.; Lee, G.; Teicholz, P. *BIM Handbook: A Guide to Building Information Modeling for Owners, Designers, Engineers, Contractors, and Facility Managers*; John Wiley Sons: Hoboken, NJ, USA, 2018.
116. Dallasega, P.; Rauch, E.; Linder, C. Industry 4.0 as an enabler of proximity for construction supply chains: A systematic literature review. *Comput. Ind.* **2018**, *99*, 205–225. [CrossRef]
117. Dave, B.; Buda, A.; Nurminen, A.; Främling, K. A framework for integrating BIM and IoT through open standards. *Autom. Constr.* **2018**, *95*, 35–45. [CrossRef]
118. Hosseini, S.M.; Carli, R.; Dotoli, M. Robust optimal energy management of a residential microgrid under uncertainties on demand and renewable power generation. *IEEE Trans. Autom. Sci. Eng.* **2020**, 1–20. [CrossRef]
119. Li, C.T.; Cheng, J.C.; Chen, K. Top 10 technologies for indoor positioning on construction sites. *Autom. Constr.* **2020**, *118*, 103309. [CrossRef]
120. Blaauwbroek, N.; Nguyen, P.H.; Konsman, M.J.; Shi, H.; Kamphuis, R.I.; Kling, W.L. Decentralized resource allocation and load scheduling for multicommodity smart energy systems. *IEEE Trans. Sustain. Energy* **2015**, *6*, 1506–1514. [CrossRef]
121. Maskuriy, R.; Selamat, A.; Ali, K.N.; Maresova, P.; Krejcar, O. Industry 4.0 for the construction industry—How ready is the industry? *Appl. Sci.* **2019**, *9*, 2819. [CrossRef]
122. Boje, C.; Guerriero, A.; Kubicki, S.; Rezgui, Y. Towards a semantic Construction Digital Twin: Directions for future research. *Autom. Constr.* **2020**, *114*, 103179. [CrossRef]
123. Lavy, S.; Jawadekar, S. A case study of using BIM and COBie for facility management. *Int. J. Facil. Manag.* **2014**, *5*, 109458840.
124. Koch, C.; Hansen, G.K.; Jacobsen, K. Missed opportunities: Two case studies of digitalization of FM in hospitals. *Facilities* **2019**, *37*, 381–394. [CrossRef]
125. Shirowzhan, S.; Sepasgozar, S.M.; Edwards, D.J.; Li, H.; Wang, C. BIM compatibility and its differentiation with interoperability challenges as an innovation factor. *Autom. Constr.* **2020**, *112*, 103086. [CrossRef]
126. Son, H.; Kim, C.; Turkan, Y. Scan-to-BIM-an overview of the current state of the art and a look ahead. In *ISARC. Proceedings of the International Symposium on Automation and Robotics in Construction*; IAARC Publications: London, UK, 2015; Volume 32, p. 1.
127. Hess, M.; Petrovic, V.; Yeager, M.; Kuester, F. Terrestrial laser scanning for the comprehensive structural health assessment of the Baptistery di San Giovanni in Florence, Italy: An integrative methodology for repeatable data acquisition, visualization and analysis. *Struct. Infrastruct. Eng.* **2018**, *14*, 247–263. [CrossRef]
128. Fricker, P.; Sandau, R.; Walker, A.S. Development of an airborne digital sensor for photogrammetric and remote sensing applications. In Proceedings of the ASPRS Annual Conference, Washington, DC, USA, 21–26 May 2000.
129. Dai, F.; Feng, Y.; Hough, R. Photogrammetric error sources and impacts on modeling and surveying in construction engineering applications. *Vis. Eng.* **2014**, *2*, 1–14. [CrossRef]
130. Tang, P.; Huber, D.; Akinci, B.; Lipman, R.; Lytle, A. Automatic reconstruction of as-built building information models from laser-scanned point clouds: A review of related techniques. *Autom. Constr.* **2010**, *19*, 829–843. [CrossRef]
131. Hernandez-Marin, S.; Wallace, A.M.; Gibson, G.J. Bayesian analysis of lidar signals with multiple returns. *IEEE Trans. Pattern Anal. Mach. Intell.* **2007**, *29*, 2170–2180. [CrossRef] [PubMed]
132. Maser, K.R. Condition assessment of transportation infrastructure using ground-penetrating radar. *J. Infrastruct. Syst.* **2007**, *2*, 94–101. [CrossRef]
133. Maierhofer, C. Nondestructive evaluation of concrete infrastructure with ground penetrating radar. *J. Mater. Civ. Eng.* **2003**, *15*, 287–297. [CrossRef]
134. Solla, M.; Gonçalves, L.M.S.; Gonçalves, G.; Francisco, C.; Puente, I.; Providência, P.; Gaspar, F.; Rodrigues, H. A Building Information Modeling Approach to Integrate Geomatic Data for the Documentation and Preservation of Cultural Heritage. *Remote Sens.* **2020**, *12*, 4028. [CrossRef]
135. Isikdag, U. Enhanced building information models. In *Enhanced Building Information Models: Using IoT Services and Integration Patterns*; Springer International Publishing : New York, NY, USA, 2015; pp. 13–24. [CrossRef]
136. Thu Nguyen, Integration of BIM and IoT to Improve Building Performance for OCCUPANTS' perspectives. Master's Thesis, KTH Royal Institute of Technology, Stockholm, Sweden, 2016.
137. Rowley, J. The wisdom hierarchy: Representations of the DIKW hierarchy. *J. Inf. Sci.* **2007**, *33*, 163–180. [CrossRef]
138. Ufuk-Gökçe, H.; Umut-Gökçe, K. Integrated system platform for energy efficient building operations. *J. Comput. Civ. Eng.* **2014**, *28*, 05014005. [CrossRef]
139. Pishdad-Bozorgi, P.; Gao, X.; Eastman, C.; Self, A.P. Planning and developing facility management-enabled building information model (FM-enabled BIM). *Autom. Constr.* **2018**, *87*, 22–38. [CrossRef]

140. Khalid, M.U.; Bashir, M.K.; Newport, D. Development of a building information modelling (BIM)-based real-time data integration system using a building management system (BMS). In *Building Information Modelling, Building Performance, Design and Smart Construction*; Springer: Cham, Switzerland, 2017; pp. 93–104.
141. Desogus, G.; Quaquero, E.; Sanna, A.; Gatto, G.; Tagliabue, L.C.; Rinaldi, S.; Ciribini, A.L.C.; Di Giuda, G.; Villa, V. Preliminary performance monitoring plan for energy retrofit: A cognitive building: The "Mandolesi Pavillon" at the University of Cagliari. In Proceedings of the 2017 AEIT International Annual Conference, Cagliari, Italy, 20–22 September 2017; pp. 1–6.
142. Hefnawy, A.; Bouras, A.; Cherifi, C. IoT for smart city services: Lifecycle approach. In Proceedings of the International Conference on Internet of things and Cloud Computing, Cambridge, UK, 23 February–22 March 2016; pp. 1–9.
143. Shabanzadeh, M.; Sheikh-El-Eslami, M.K.; Haghifam, M.R. A medium-term coalition-forming model of heterogeneous DERs for a commercial virtual power plant. *Appl. Energy* **2016**, *169*, 663–681. [CrossRef]
144. Kanan, R.; Elhassan, O.; Bensalem, R. An IoT-based autonomous system for workers' safety in construction sites with real-time alarming, monitoring, and positioning strategies. *Autom. Constr.* **2018**, *88*, 73–86. [CrossRef]

MDPI
St. Alban-Anlage 66
4052 Basel
Switzerland
Tel. +41 61 683 77 34
Fax +41 61 302 89 18
www.mdpi.com

Sustainability Editorial Office
E-mail: sustainability@mdpi.com
www.mdpi.com/journal/sustainability

www.ingramcontent.com/pod-product-compliance
Lightning Source LLC
LaVergne TN
LVHW070928170726
843515LV00005B/899

* 9 7 8 3 0 3 6 5 5 5 1 9 5 *